Orestes M. Brands

Lessons on the Human Body

Orestes M. Brands

Lessons on the Human Body

ISBN/EAN: 9783337371586

Printed in Europe, USA, Canada, Australia, Japan

Cover: Foto ©berggeist007 / pixelio.de

More available books at **www.hansebooks.com**

LESSONS

ON

THE HUMAN BODY.

An Elementary Treatise

UPON PHYSIOLOGY, HYGIENE, AND THE EFFECTS
OF STIMULANTS AND NARCOTICS ON
THE HUMAN SYSTEM.

BY

ORESTES M. BRANDS,

PRINCIPAL OF GRAMMAR AND·PRIMARY SCHOOL NO. 4,
PATERSON, N.J.

LEACH, SHEWELL, & SANBORN,
BOSTON AND NEW YORK.

PREFACE.

THE formidable size and ponderous character of many books placed in the hands of children have been prolific sources of discouragement of effort, and, not infrequently, causes for dislike and neglect of important and interesting studies.

These simple Lessons on the Human Body are specially designed to present subject-matter in such quantity and of such quality as shall make it *possible and probable* that the young student may " make its acquaintance."

It is confidently believed that the arrangement of the material will at once commend itself to the teacher. Attention is respectfully directed to the following features; viz., —

1. Short, complete lessons.
2. The systematic division of each lesson that describes an organ into three distinct topics, — *Position, Construction, Work.*
3. The arrangement of the entire text in short, numbered paragraphs, each stating an important fact briefly.

4. The adaptation of the text to oral instruction, *the teacher's work being already arranged.*

While in manuscript form, these lessons were used, with much success, in large schools.

Thanks are due to LeRoy F. Lewis, Principal of School No. 11, Brooklyn, who unites with his high qualifications as a teacher special scientific and professional knowledge, for valuable suggestions; and to Dr. Albert Day of the Washingtonian Home, Boston, an eminent writer and authority on alcoholic diseases, who has read the manuscript on alcohol, and gives it his unqualified approval.

The author believes it to be unnecessary to waste time and space in presenting the importance of an *early* acquaintance with the structure and functions of the principal organs of the human body, and of a general knowledge of the laws governing their well-being. No intelligent person of to-day questions the importance of such knowledge. If this little book should merit the approbation of my fellow-teachers, I shall feel fully repaid for the labor attending its preparation.

O. M. B.

Dec. 22, 1883.

CONTENTS.

PART I.

THE SKELETON.

Contents.

PART II.

DIGESTION.

PART III.

THE BLOOD AND ITS CIRCULATION.

Contents.

PART IV.

THE BREATHING APPARATUS.

PART V.

THE MUSCLES.

PART VI.

THE BRAIN AND NERVES.

PART VII.

EYE, EAR, AND SKIN.

PART VIII.

ALCOHOL AND THE HUMAN SYSTEM.

Contents.

xii   *Contents.*

PART IX.

TOBACCO.

APPENDIX.

PART I.

THE SKELETON.

"Knowest thou the nature of the human frame,
That world of wonders more than we can name?
Say, has thy busy, curious eye surveyed
The proofs of wisdom here displayed?"

THE SKELETON.

Lesson I.

(a) *Definition.* — **1.** The skeleton is the framework of the body.

(b) *Number of the Bones.* — **1.** The skeleton is composed of about 208 bones. The number varies at different periods of life. What is merely gristle in infancy becomes bone later in life.

(c) *Uses of the Bones.* — **1.** Some bones protect the delicate organs enclosed by them.

2. Many of the bones give shape to and preserve the form of the body.

3. A large number of the bones serve as levers, on which the muscles may act to produce motion.

(d) *Form of the Bones.* — **1.** Some bones are long, as in the legs, for convenience in walking; and hollow, to give lightness.

2. Where much strength in small space is needed, the bones are short and thick.

3. Bones that cover cavities are broad and flat, as in the chest and skull.

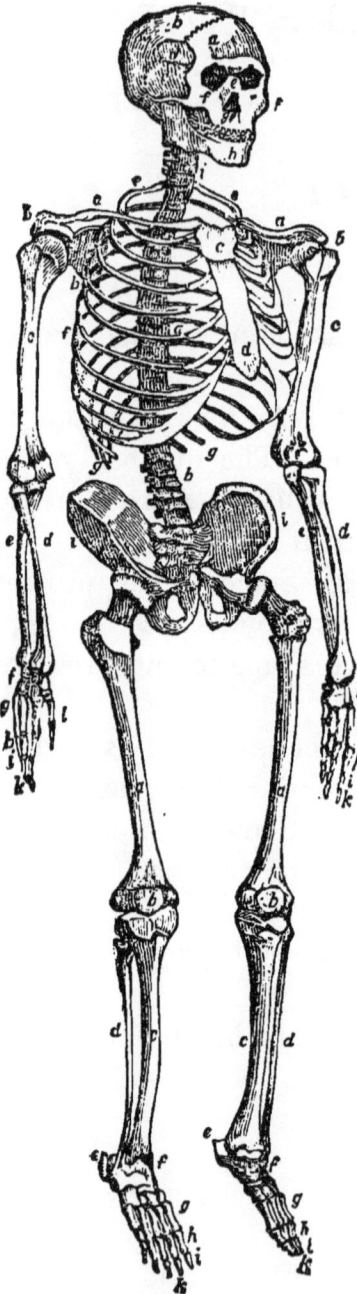

A Front View of the Male Skeleton.

Head and Neck.

a, the frontal bone.
b, the parietal bone.
c, the temporal bone.
d, a portion of the sphenoid bone.
e, the nasal bone.
f, the malar, or cheek-bones.
g, the superior maxillary, or upper jaw.
h, the lower jaw.
  i, the bones of the neck.

Trunk.

a, the twelve bones of the back.
b, the five bones of the loins.
c, *d*, the breast-bone.
e, *f*, the seven true ribs.
g, *g*, the five false ribs.
h, the rump-bone, or sacrum.
i, The hip-bones.

Upper Extremity.

a, the collar-bone.
b, the shoulder-blade.
c, the upper-arm bone.
d, the radius.
e, the ulna.
f, the carpus, or wrist.
g, the bones of the hand.
h, first row of finger-bones.
  i, second row of finger-bones.
k, third row of finger-bones.
l, the bones of the thumb.

Lower Extremity.

a, the thigh-bone.
b, the knee-pan.
c, the tibia, or large bone of the leg.
d, the fibula, or small bone of the leg.
e, the heel-bone.
f, the bones of the instep.
g, the bones of the foot.
h, the first row of toe-bones.
  i, the second row of toe-bones.
k, the third row of toe-bones.

Fig. 1.

A BACK VIEW OF THE MALE
SKELETON.

The Head.

a, the parietal bone.
b, the occipital bone.
c, the temporal bone.
d, the cheek-bone.
e, the lower jaw-bone.

Neck and Trunk.

a, the bones of the neck.
b, the bones of the back.
c, the bones of the loins.
d, the hip-bone.
e, the sacrum.

Upper Extremity.

a, the collar-bone.
b, the blade-bone.
c, the upper bone of the arm.
d, the radius.
e, the ulna.
f, the bones of the wrist.
g, the bones of the hand.
h, the first row of finger-bones.
i, the second row of finger-bones.
k, the third row of finger-bones.
l, the bones of the thumb.

Lower Extremity.

a, the thigh-bone.
b, the large bone of the leg.
c, the small bone of the leg.
d, the heel-bone.
e, the bones of the instep.
f, the bones of the toes.

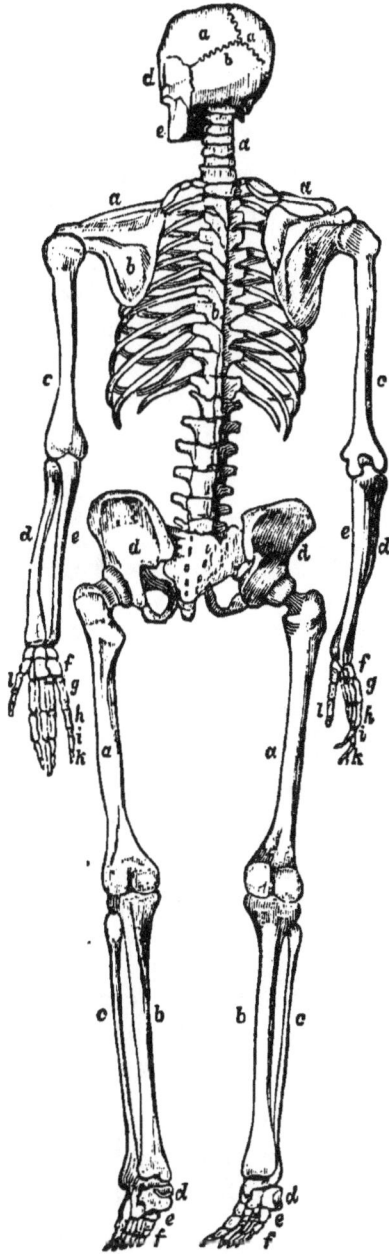

FIG. 2.

4. A large number of bones are irregular in shape, to suit particular purposes.

5. The general form of the bones is such as gives firmness and strength without great weight.

Remarks. — If the bones of the limbs were solid, they would be much heavier, and therefore not so well adapted to rapid movement. Their hollow form gives them greater strength than the same amount of bone would have in a solid form.

Lesson II.

THE BONES IN GENERAL. — *Continued.*

(a) *Composition.* — **1.** The bones are composed of *animal matter*, or jelly, and of *mineral matter*, — lime, etc.

(b) *Use of the Materials.* — **1.** The mineral matter gives hardness and stiffness to the bones.

2. The animal matter gives toughness and elasticity.

(c) *Structure of the Bones.* — **1.** The bones are hard externally, but are somewhat softer, and hollow, within.

2. The hollow portions are filled with a spongy substance composed of marrow and blood-vessels.

3. In infancy the bones are only cartilage; but this gradually hardens by additions of mineral matter, and in a few years becomes firm bone. In early life

the bones are so tough as not to be easily broken;
but in old age the greater amount of earthy matter
in them causes brittleness, and when broken they do
not heal so quickly as in youth.

Lesson III.

THE BONES IN GENERAL. — Concluded.

(a) *Growth of the Bones.*—1. Bone once formed
does not remain during life, but is constantly disap-
pearing and being renewed in all its parts, gradually, but continually.

2. The growth of a bone, as a general rule, takes
place only by addition to its free ends and surfaces.

3. The blood circulates freely through the bones,
and supplies them with materials required for their
growth and nourishment.

(b) *Repair of Broken Bone.*—1. Nature has a
process of her own in repairing broken bones. As
soon as she can check the flow of blood from the
broken ends, she sends out a watery fluid which
contains material of which gristle is formed. In a
few days the gristle becomes tough, and holds the
bones in place till mineral can be added to complete
the union of the broken part. A length of time is
required to complete the firm repair, and great care
is required in the use of the bone in the mean time.

Remarks. — In none of the organs of the body is the constant change of particles which compose them so easily noticed as in the bones. If we mix madder with the food of an animal, the bones soon become red, and they regain their original color when the coloring-matter no longer forms part of the food. Again: if the madder be given for a time, and then omitted, and after a while given again, the bones show a white streak between two red ones; which proves that they grow from the surface toward the centre.

Nature gives additional strength to the broken bone by forming a ring, or ridge of bone, at the place where it has been broken.

Lesson IV.

JOINTS.

(a) *Position.* — **1.** Bones are connected at their ends or at their sides.

2. The point at which the bones are connected is a *joint.*

(b) *Construction.* — **1.** The ends of the bones forming a joint are covered with a thick and somewhat elastic cartilage, or gristle.

2. The cartilage is again covered with a thin substance, the *synovial membrane,* which gives out a fluid like the white of an egg. This oils the joints, so that the bones may move freely.

3. The bones forming the joints are held together by strong cords or bands of gristle called *ligaments* (from *ligo*, to bind).

(c) *Work of the Joints.* — 1. The joints permit the bones to move, and change position; so that the limbs and other portions of the body may bend, and thus perform the various offices that may be required of them.

—————

Lesson V.

JOINTS. — *Concluded.*

(a) *Kinds.* — 1. There are two kinds of movable joints; viz., the *hinge-joint* and the *ball-and-socket joint.*

(b) *Construction.* — 1. The hinge-joint is so constructed as to permit motion in only one direction, as that of the elbow.

2. The ball-and-socket joint is so formed as to allow motion in every direction, — forward and backward, and in a circular manner. It is composed of a ball on the end of one bone, and a cup or socket in another, into which the ball fits. The shoulder-joint is of this kind.

Remarks. — The animal body is the only machine that makes the oil which lubricates its own joints. The *synovial fluid* is the oil of the joints.

The joints, though in such frequent use from infancy to old age, seldom wear out. The tough covering of the ends of the bones is as thick and smooth at the end of life as at the beginning.

Lesson VI.

CLASSES OF BONES. — BONES OF THE HEAD.

The bones of the skeleton are divided into four classes; viz., —

1. Bones of the Head.

2. Bones of the Trunk.

3. Bones of the Upper Extremities.

4. Bones of the Lower Extremities.

Bones of the Head.

(a) *Location.* — **1.** There are thirty bones in the head, and they are located as follows: —

. **2.** Skull, 8 bones.

Face, 14 bones.

Ears, 8 bones.

Besides these, there are thirty-two teeth.

(b) *Construction.* — **1.** The bones of the skull form a hollow, or cavity, in which the brain is situated.

2. The bones of the skull are united by a sort of notched joint, similar to what carpenters name " dovetailed " joint. These joints are called *sutures.*

3. The form of the skull is oval, and is adapted to resist pressure.

4. The front of the skull is narrower and stronger than the back, and is thus prepared to protect the brain at a point where danger is greatest.

5. The elastic packing between the bones, at the joints, prevents much of the jar from blows.

6. All the bones of the head, excepting the lower jaw, are immovable.

EXPLANATIONS OF FIG. 3.

a, a, the *coronal suture.*
b, the *sagittal suture.*
c, the *lambdoidal suture.*
d, d, ossa triquetra, small ragged bones, occasionally found in some skulls, lying in the last-mentioned suture.
e, e, portions of the *temporal bone,* overlapping the walls.

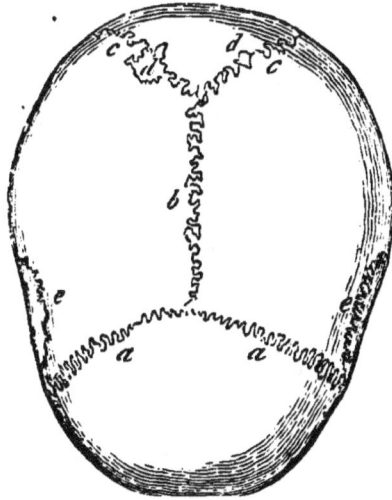

FIG. 3.

(c) *Work.* — **1.** The bones of the skull and face, protect the organs of sense — smell, taste, hearing, and sight — from injury.

2. The bones of the ear aid in hearing.

3. The bones of the lower jaw are provided with hinge-joints, so as to permit the opening and closing of the mouth, the movements required in masticating food, etc.

4. The teeth are used in cutting and grinding the food.

Lesson VII.

BONES OF THE TRUNK.

(a) *Location.* — **1.** The trunk is that portion of the body situated between the upper and the lower extremities. It contains fifty-four bones, located as follows; viz. (*see Figs. 1 and 2*), —

The Spine, 24 bones.
The Ribs, 24 bones.
The Pelvis, 4 bones.
The Sternum, 1 bone.
Root of Tongue, 1 bone.

(b) *Structure.* — **1.** The trunk contains two cavities enclosed by the ribs, sternum, spine, and bones of the pelvis.

2. The upper cavity, the *chest*, contains the heart, lungs, etc.

3. The lower cavity, the *abdomen*, contains the stomach, liver, kidneys, and the intestines.

4. The two cavities are separated by a muscular partition called the *diaphragm*, and are enclosed by the ribs and muscular walls of the abdomen.

Remarks. — The diaphragm is a great partition situated between the chest and the abdomen, having its convex or rounded surface toward the chest, and its concave or hollowed side toward the abdomen. Its work is explained in the lessons on the Breathing Apparatus.

Lesson VIII.

THE THORAX, OR CHEST.

(a) *Position.* — 1. The *thorax*, or chest, is the upper and smaller of the two great cavities of the trunk.

EXPLANATION OF FIG. 4.

This figure represents the *sternum*, or breast-bone.

A, the place where the collar-bone is joined.

C, where the first rib is joined.

c, d, e, f, g, the number of pieces which are united into one.

h, the tip of the sternum.

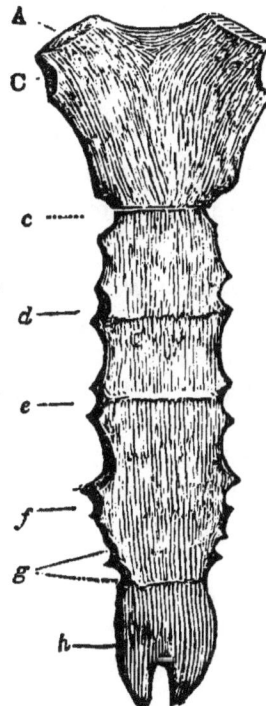

FIG. 4.

(b) *Construction.* — 1. The natural form of the chest is that of a cone diminishing upward, its apex being between the shoulders. (*See Fig. 1.*)

2. Its peculiar construction allows a great variety of bodily movement, — bending and straightening the trunk, movements from side to side, and also a rotary movement, enabling us to twist the trunk nearly one-fourth of a turn, — thus permitting the great number of movements required for convenience in labor, pleasure, etc.

3. The spine is a wonderful piece of mechanism. In it we have a column of twenty-four bones, united so ingeniously and firmly as to sustain a heavy load, and yet so elastic that it will bend like rubber, keeping the body proudly and sturdily erect when we will, or permitting it to bend in humble obedience to our inclinations.

Remarks. — The pads of cartilage between the bones of the spine vary from one-fourth to one-half of an inch in thickness. They become compressed by the weight they bear during . the day; so that a man is not quite as tall in the evening as in the morning; but, as the pads are elastic, they recover their thickness during the night, or when pressure upon them is removed. A man is somewhat shorter in old age than at earlier periods of his life, because long-continued pressure of the weight of the head and upper parts of the body, together with the burdens of labor, overcome the elasticity of the pads, and they remain thin or compressed. The backbone thus becomes slightly shortened.

Lesson X.

THE RIBS.

(a) *Position.*—1. The *ribs* are slender, curved bones, arranged in pairs, twelve on each side of the chest. (*See Figs. 1 and 2.*)

(b) *Construction.*—1. The ribs are attached by their heads to the spine; and by means of cartilage their other extremities are attached to the *sternum*, or breast-bone.

2. The seven uppermost are called *true* ribs, because each of them is connected directly with the sternum.

3. The five lower ribs are called *false* ribs, because one or two of them are loose at one end, and the others run together, instead of being separately extended to and connected with the breast-bone.

(c) *Work.*—1. The use of the ribs is to form the cavity of the chest for the reception and protection of the lungs, heart, and great blood-vessels.

2. The ribs also assist in breathing by their alternate rising and falling. This action enlarges and diminishes the size of the chest, giving space for the expansion of the lungs.

3. The slenderness and curved form of the ribs give lightness and strength, while the elastic cartilages permit freedom of movement. Here, as elsewhere, Nature has provided what is required to carry on her work safely and freely.

Lesson XI.

THE PELVIS.

(a) **Position.** — **1.** The *pelvis* is the bony structure at the base of the trunk. (*See Fig. 1.*)

(b) **Construction.** — **1.** The pelvis is composed of three bones, — the two hip-bones and the *sacrum*, a wedge-shaped bone situated between the hip-bones.

2. These bones are broad and flat, and are spread out to form a sort of basin, on which the abdomen rests. The spine stands on the sacrum, and the thigh-bones are attached to the hip-bones.

(c) **Work.** — **1.** The office of the pelvis is to provide a strong foundation for the support of the bones of the spine and for the weight of the body above it.

2. The pelvis also furnishes sockets for the attachment of the thigh-bones.

Remarks. — The hip-bones are called by anatomists the *innominata*, or nameless bones.

The *sacrum* (sacred), so called because it was anciently offered in sacrifice, stands, like the keystone of an arch, between the innominata, or hip-bones.

Lesson XII.

BONES OF THE UPPER EXTREMITIES.

(a) The bones of the upper extremities are, —
1. Collar-bone (*clavicle*) . . . . . . 2 bones.
2. Shoulder-blade (*scapula*) . . . . . 2 bones.
3. Bones of upper arm (*humerus*) . . 2 bones.
4. Bones of lower arm (*ulna and radius*), 4 bones.
5. Bones of wrist (*carpus*) . . . . . 16 bones.
6. Bones of hand (*metacarpus*) . . . 10 bones.
7. Bones of fingers (*phalanges*) . . . 28 bones.
    Total . . . . 64 bones.

Lesson XIII.

THE SCAPULA.

(a) *Position.* — 1. The *scapula* lies at the top and back of the chest, and is familiarly known as the *shoulder-blade.* (*See Fig. 2.*)

(b) *Construction.* — 1. The scapula is a broad, thin, flat, triangular bone embedded in the flesh, and held in its place by muscles.

2. It is not directly attached to the trunk.

3. At its upper and outer corner it is connected with the collar-bone (*clavicle*), and at this point it

has a shallow socket for the head of the bone of the upper arm (the *humerus*).

(c) *Work.* — 1. The scapula affords a foundation for the attachment of the muscles of the shoulders.

2. The scapula also aids in forming the shoulder-joint, serving to connect the arm with the trunk of the body.

————

Lesson XIV.

THE CLAVICLE, OR COLLAR-BONE.

(a) *Location.* — 1. The *clavicle* is located at the top and in front of the chest.

(b) *Construction.* — 1. The clavicle (*clavis*, a key) is a long, slender bone, shaped like the Italic *f*.

2. It is fastened at one end to the breast-bone and the first rib, and at the other to the shoulder-blade. (*See Fig. 1.*)

(c) *Work.* — 1. The clavicle acts as a brace to hold the shoulder-joint out from the chest, and thus gives the arm greater play.

Remarks. — If the clavicle be removed or broken, the head of the arm-bone will fall, and the motions of the arm be greatly restricted.

The lower animals, whose front limbs are near each other, have no collar-bone.

Lesson XV.

THE SHOULDER-JOINT.

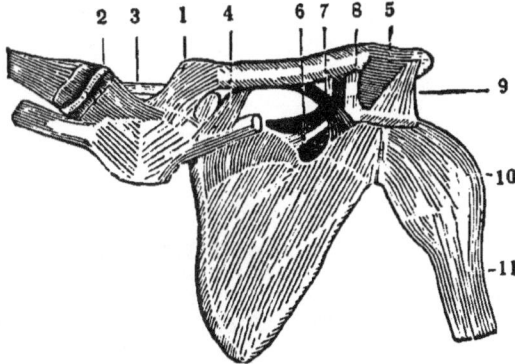

EXPLANATION OF FIG. 6.

In this cut is seen the union of the shoulder-blade, collar-bone, breast-bone, and the shoulder-joint. These are detached from the body: hence the view is a front one. A portion of the collar-bone of the right side is seen also, all the others being on the left side. The figures from 1 to 11 indicate the ligaments which keep them united when the muscles are dissected away.

FIG. 6.

(a) *Location.* — 1. The shoulder-joint is located at the junction of the scapula and the humerus (bone of the upper arm).

(b) *Construction.* — 1. The humerus articulates (joins with) the scapula, and forms a ball-and-socket joint.

2. This joint consists of a shallow, cup-like cavity in the scapula, into which the rounded head of the humerus fits.

(c) *Work.* — 1. The shoulder-joint permits a free, rotary motion, allowing the arm to move in any direction.

Remarks. — The shoulder-joint is easily dislocated (put " out of joint "), because its socket is so shallow: still, if it were deeper, the arm could not move so freely.

Lesson XVI.

THE ARMS.

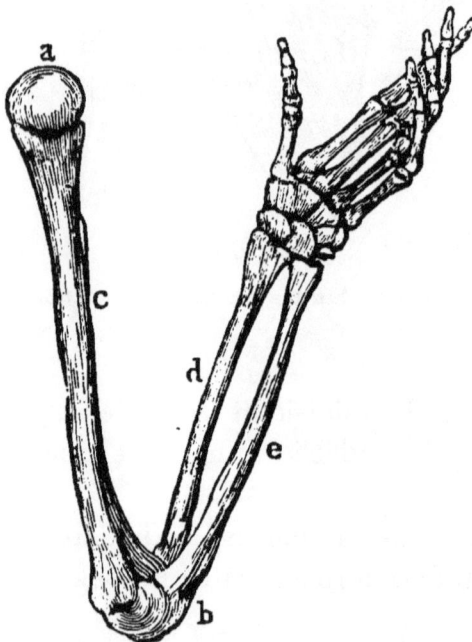

FIG. 7.

EXPLANATION OF
FIG. 7.

All the bones of the arm, fore-arm, and hand, are here exhibited in connection, with reference to impressing it on the mind, after having read a short description of the individual parts of the upper extremity.

a is the head of the arm-bone, articulated to the shoulder.

b, the joint, or elbow, formed by the *ulna* and lower end of the arm.

c, the shaft of the *os humeri*, or arm.

d, the *radius*, or handle of the hand, united solely to the wrist.

e, the *ulna*, which alone forms with the arm the joint.

(a) *Construction.* — 1. That portion of an arm between the shoulder and the elbow consists of a single bone, called the *humerus.*

2. That portion of an arm between the elbow and

the wrist is composed of two bones, the *ulna* and the *radius.*

3. The radius extends to the hand; but the ulna, while connected with the elbow, does not reach the hand.

4. The *ulna* is the smaller of the two bones of the lower part of the arm, and it is situated on the inner or little-finger side of the arm. The *radius* is placed on the outer or thumb side of the arm.

5. The arms are attached to and suspended from the scapula, at the shoulder.

6. The bones of the arm furnish attachment for a large number of muscles that move the hand and fingers.

Lesson XVII.

THE ELBOW.

(a) *Position.* — **1.** The lower end of the *humerus* articulates with the upper ends of the ulna and radius, forming a hinge-joint known as the *elbow.*

(b) *Construction.* — **1.** At the elbow, the rounded head of the radius fits into a shallow cavity in the ulna.

2. The ulna at the elbow is large, and it assists in giving strength to the joint.

(c) *Work.* — **1.** The upper end of the radius turns upon the double surface furnished it by the ball of

the humerus and the partial cup of the ulna, allowing a gliding motion in such a way that the palm of
the hand may turn in different directions.

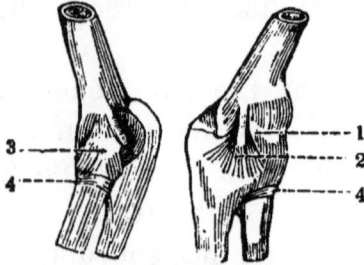

FIG. 8.

EXPLANATION OF FIG. 8.

Short ligaments of the elbow are here
demonstrated. The wonder is, how the
elbow-joint can ever be dislocated without entirely ruining the whole ligamen-
tary arrangement. The figures from 1 to
4 not only give the locality of each ligament, but even the figure.

2. The elbow-joint permits motion in two ways,
i.e., backward and forward, and a rotary motion of
the lower arm.

Lesson XVIII.

THE WRIST.

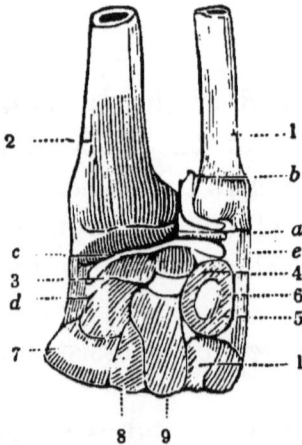

FIG. 9.

EXPLANATION OF FIG. 9.

This diagram shows the connection of the
little bones of the *carpus*, or wrist, with the two
long bones of the fore-arm.

1, the *ulna.*
2, *radius.*
3, *scaphoides.*
4, *lunare.*
5, *cuneiforme.*
6, *pisiforme.*
7, *trapezium.*
8, *trapeziodes.*
9, *magnum.*

The letters mark the ligaments which tie
them together.

(a) **Position.** — 1. The *wrist* is located between the arm and the hand.

(b) **Construction.** — 1. The wrist, or *carpus*, consists of eight very irregular bones, arranged in two rows.

2. One of these rows articulates with the bones of the arm ; the other, with the bones of the hand.

EXPLANATION OF FIG. 10.

Another plan of the bones of the wrist, showing them placed in two rows. This is a back view of the carpus of the right hand.

a, the *boat-shaped bone ;*
b, the *half-moon shaped ;*
c, the *wedge-shaped ;*
d, the *pea-shaped ;* which make the upper row, joining the fore-arm.

In the second row are the four others, e, f, g, h, which are united by a joint to the palm of the hand.

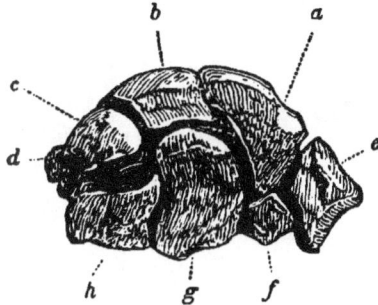

FIG. 10.

3. The bones are held so firmly together by ligaments that they are seldom displaced.

(c) **Work.** — 1. The wrist forms a hinge-joint, and admits of motion in two directions, *i.e.*, backward and forward, and a gliding motion from side to side.

2. The arrangement of these bones admits of but little variety of motion, but combines great strength with elasticity.

Lesson XIX.

THE HAND.

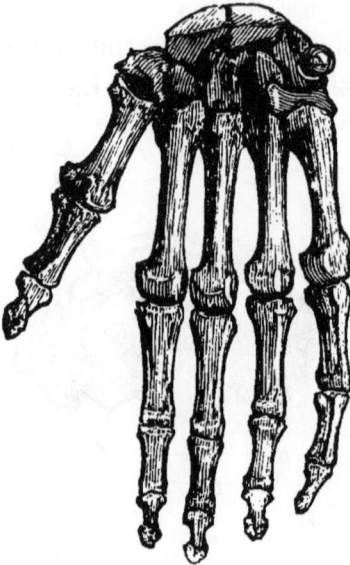

EXPLANATION OF FIG. 11.

Here is presented a back view of all
the bones of the hand as they are connected with the eight little bones of the
wrist. Each bone is so distinctly represented, that a very young child may
understand the arrangement.

FIG. 11.

(a) *Position.*—1. The bones of the palm of the
hand, *metacarpus* (*meta*, beyond; and *karpos*, wrist),
articulate with the bones of the wrist. The metacarpal bones are five in number in each hand.

2. Each of the bones of the palm articulates with
a thumb or a finger, the bones of which are named
phalanges (the plural of *phalanx*, meaning a rank).
The metacarpus and phalanges comprise the bones
of the hand.

(b) *Construction.*—1. The bones of the palm articulate at one end with the bones of the wrist, and at the other with the bones of the fingers.

2. The first bones of the fingers are so joined to the palm of the hand as to permit the motion of a hinge-joint, and also of a sidewise motion. The other bones of the fingers form simple hinge-joints.

3. The first bones of the thumbs are not connected with the others of the fingers, and have a freedom of motion peculiar to themselves.

4. There are three bones in each finger, and but two in each thumb.

(c) *Work.*—**1.** The hand is beautifully and skilfully arranged, and adapted to an almost infinite variety of purposes.

2. The numerous joints of the fingers, and the varying length of their bones, enable them to fit the hollow of the hand when it is closed, and to grasp objects of varying size, from a fine needle to a large bar of iron.

Remarks.—The hand in its perfection belongs to man alone. Its wonderful structure is suited to obey the requirements of the mind which directs it, and gives to man a superiority over all other animals; for none other is equipped with an instrument so fully capable of performing the great variety of motion and work.

The hand is not only a wonderful instrument of motion, but it is also the chief organ of touch or feeling. And what a delicate instrument it is for this purpose!

Lesson XX.

BONES OF THE LOWER EXTREMITIES.

a. The bones of the lower extremities are, viz.,[1] —

1. The *femur*, or thigh-bone . . . . 2 bones.

2. The *patella*, or knee-pan . . . . . 2 bones.

3. The *tibia*, or shin-bone . . . . . 2 bones.

4. The *fibula*, or smaller bone of the
lower leg . . . . 2 bones.

5. The *tarsals*, or bones of the instep . 14 bones.

6. The *metatarsals*, or bones beyond the
instep . . . . 10 bones.

7. The *phalanges*, or bones of the toes . 28 bones.

Total . . . . 60 bones.

Lesson XXI.

THE HIP AND KNEE.

(a) *Structure of the Hip-Joint.* — **1.** The *femur* (*femoris*, the thigh), or thigh-bone, articulates with the hip-bone (pelvis), and forms a ball-and-socket joint.

2. In the sides of the hip-bones there are cup-like hollows, into which the upper end of the femur fits snugly. A strong ligament, attached to the ball-like

---

[1] The number of bones given above includes both of the legs.

end of the femur and to the centre of the socket, binds the bones together.

3. So tightly does the femur[1] fit in the deep socket, that .the pressure of the air holds it in place, even after the flesh is removed, and considerable force is required to separate the ball from the socket.

EXPLANATION OF FIG. 12.

This is a drawing of the lower part of the hip-bone, or *os innominatum*, in which is seen the head of the thigh-bone, tied into its socket by a short round cord, to keep it always in place. Were it not for this curious provision, by a thousand unguarded movements the hip would be thrown out of joint.

b, the *cord* that keeps the bone in its socket.

c, the socket in the hip-bone.

d, a rim of the socket, to deepen it.

f, the thigh-bone head.

6, the point of bone on which we sit.

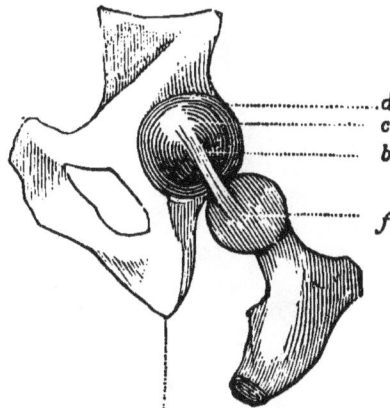

FIG. 12.

(b) *Work.* — **1.** The hip-joint permits the raising of the leg, as in walking.

(c) *Construction of the Knee.* — **1.** The lower end of the *femur* joins the upper end of the *tibia*, and forms the hinge-joint known as the knee-joint.

2. The *patella* (*patina*, a little dish), a chestnut-shaped bone, is firmly fastened over the joint in front. It protects and strengthens the joint.

---

[1] The femur is the longest and strongest bone of the body. It bears the entire weight of the parts above it at every step.

3. The *fibula* (*fibula*, a clasp), the small outer bone of the lower leg, is securely fastened at both ends to the shin-bone. Its lower end may be felt on the outer ankle. This bone does not form a part of the joints, but seems merely to brace the tibia, and to offer a place for the attachment of muscles. It probably protects the ankle-joint.

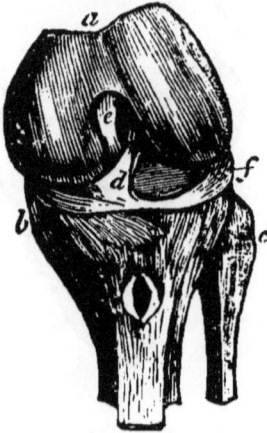

EXPLANATION OF FIG. 13.

e, d, are the *crucials*, or cross ligaments, remarkable in structure and office.

f, the tendon of an extensor muscle.

c, the head of the *fibula*, joining the side of the shin-bone.

a, the articulating surface of the lower end of the thigh-bone, covered by the knee-pan.

b refers to the broad ligament, turned down from the joint to expose the cross ligaments, having the knee-pan on it.

FIG. 13.

(d) *Work.* — **1.** The knee-joint permits flexion, or bending of the limb, about midway of its length, in a direction opposite that provided by the hip-joint. It also allows a slight rotary motion.

Remarks. — The patella, being held in its place only by muscles, is easily displaced, and frequently slips aside. From its position it is extremely liable to receive blows which would otherwise fall directly upon the other bones of the joint; and, while protecting these, it not infrequently becomes fractured. Such an injury, however, is not so serious as the fracture of the other bones would be.

Lesson XXII.

THE ANKLE AND FOOT.

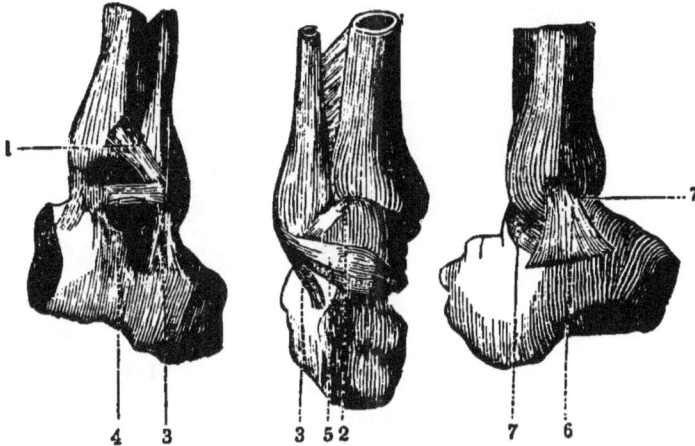

FIG. 14.

EXPLANATION OF FIG. 14.

These three plans show how the two bones of the leg are united above the ankle-joint. 1, 2, 3, 4, 5, 7, 7, 6, mark the ligaments which bind them firmly.

(a) *Structure of the Ankle.* — 1. The lower end of the tibia articulates with the *tarsals* (*tarsus*, the ankle), or bones of the instep, forming a hinge-joint.

(b) *Work.* — 1. The ankle-joint permits the bending necessary to easy motion of the foot in walking. Without this joint the foot could only be raised and lowered stiffly, without the rocking motion seen in walking. This joint also admits a slight sidewise or wagging movement of the foot.

(c) *Structure of the Foot.* — 1. The foot consists of twenty-six bones, — seven tarsal bones, five metatarsals, and fourteen phalanges.

2. The structure of the foot is very similar to that of the hand.

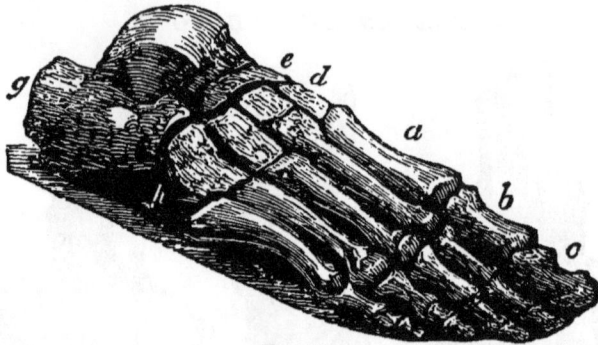

FIG. 15.

EXPLANATION OF FIG. 15.

By this diagram the skeleton of the foot will be clearly understood, even without the aid of the bones. Twenty-six bones are here so curiously grouped together, that an arch is made between the heel and ball of the great toe.

a shows the five bones of the *metatarsus.*

d, e, g, and *h* point out the five bones of the instep, or *tarsus.*

b and *c* indicate the *phalanges,* or toes.

3. The tarsal bones form the arch or instep of the foot. These bones are irregular in shape, but exactly adapted to each other. They are firmly, but not immovably, bound together by ligaments. The arch, therefore, allows a little spring to the foot, giving it elasticity.

4. The bones of the instep articulate with the *metatarsals* (*meta*, beyond, and *tarsus*, ankle), and these again each articulate with the first bone of a toe, precisely as the bones of the palm of the hand join the bones of the fingers.

(d) *Work.*—**1.** The foot is the instrument used in walking, running, and standing, and serves as a base for the support of the entire body when in an ere'ct position.

2. When the foot is not cramped by tight shoes, its action is very graceful and elastic. As we step, the weight is first thrown on the ball of the foot causing the sole to broaden and lengthen. The toes spread apart, and the springy arch of the foot aids in lowering the heel to the ground with but little jar, thus completing a step.

Remarks. — In consequence of the shoes worn by the people of civilized countries, deformity of the feet is very common. The shoes that are usually worn are narrowed in front of the ball of the foot, the toes are crowded together, — sometimes cross one another, — while in-grown nails, enlarged joints, corns, and bunions result from forcing the foot into unnatural and constrained position. Freedom and grace of movement are impossible under such conditions. Again: the extremely high-heeled shoe throws the weight of the body almost entirely upon the toes, and overtasks the muscles of the ball of the foot and calf of the leg.

Lesson XXIII.

RECAPITULATION. — CLASSES OF BONES.

(a) *Bones of the Head: —*

1. Skull . . . . 8 bones.
2. Face . . . . 14 bones.
3. Ears . . . . 8 bones.
   Total . . . . 30 bones.

(b) *Bones of the Trunk:—*
1. Spine . . . . 24 bones.
2. Ribs . . . . 24 bones.
3. Sternum . . . . 1 bone.
4. Tongue . . . . 1 bone.
5. Pelvis . . . . 4 bones.
   Total . . . . 54 bones.

(c) *Bones of the Upper Extremities:—*
1. Collar-bone (*clavicle*) . . . . . 2 bones.
2. Shoulder-blade (*scapula*) . . . . 2 bones.
3. Upper arm (*humerus*) . . . . . 2 bones.
4. Lower arm (*ulna* and *radius*) . . 4 bones.
5. Wrist (*carpus*) . . . . 16 bones.
6. Hand (*metacarpus*) . . . . . . 10 bones.
7. Fingers (*phalanges*) . . . . . . 28 bones.
   Total . . . . 64 bones.

(d) *Bones of the Lower Extremities:—*
1. Thigh-bone (*femur*) . . . . . . 2 bones.
2. Knee-pan (*patella*) . . . . . . 2 bones.
3. Shin-bone (*tibia*) . . . . 2 bones.
4. Small bones of lower leg (*fibula*) . 2 bones.
5. Instep (*tarsals*) . . . . 14 bones.
6. Beyond the instep (*metatarsals*) . 10 bones.
7. Toes (*phalanges*) . . . . . . 28 bones.
   Total . . . . 60 bones.

Aggregate number in the skeleton . 208 bones.

Lesson XXIV.

EXERCISE, DRESS, AND DEFORMITY.

(a) *Exercise.*—1. The health of the bones, as much as that of any other portion of the body, depends upon their proper nourishment and exercise.

2. When a child is feeble and unhealthy, or when it grows up without exercise, the bones do not become firm and hard as they do when healthfully developed by exercise.

3. The size and strength of the bones, to a considerable extent, depend upon exercise and good health.

(b) *Dress and Deformity.*—1. Distortion of the spine is produced by tight clothing about the waist. The liver occupies the right side of the body, while on the left side is the larger part of the stomach, which is often nearly empty. The tight clothing about the waist presses the spine over sidewise toward the unsupported part where the stomach lies: the elastic disks, or pads, between the bones are compressed on one side till they become thin, and harden into a wedge-like shape. This causes what is called *lateral curvature of the spine*, in which one shoulder becomes higher than the other.

2. Many a school-girl whose waist was originally of a proper and healthful size has gradually pressed the soft bones of youth by tight clothing at the waist, till the lower ribs, that should rise and fall with every breath, become entirely unused, and the organs of the chest and abdomen are forced out of place, distorted and hampered in their work by the compression. The troubles induced by this habit are of the most serious character. Diseases of the liver, dyspepsia, and consumption are among its legitimate results, while other disorders of a less definite nature are directly traceable to the same cause.

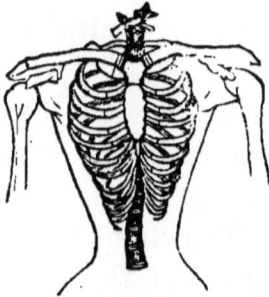

CONTRACTED CHEST.

An outline is here presented of the chest of a female, to show the condition of the bones, as they appear after death, in every woman who has habitually worn stays.

All the false ribs, from the lower end of the breast-bone, are unnaturally cramped inward towards the spine; so that the liver, stomach, and other digestive organs in the immediate vicinity, are pressed into such small compass that their functions are interrupted, and, in fact, all the vessels, bones, and viscera on which the individual is constantly depending for health, are more or less distorted and enfeebled.

FIG. 16.

3. Another distortion of the spine is produced by constant stooping of the head over books or work. This constant bending of the head forward compresses the pads, or disks, in front, while they grow thick at the back. Hardening in this shape, they act as wedges which effectually prevent, in course of time, the head from assuming an erect position, causing the awkward projection forward of the head which is so often seen. Curvature of the spine is

frequently caused by writing at desks which are too high, and which cause one shoulder to be raised higher than the other.

SKELETON OF A WELL-FORMED FEMALE CHEST.

By comparing the accompanying plan of a well-developed and naturally proportioned female chest, with the frightful skeleton appended to the preceding note, the difference is strikingly apparent. Here is breadth, space for the lungs to act in; and the short ribs are thrown outward, instead of being curved and twisted down towards the spine, by which ample space is afforded for the free action of all those organs which in the other frame were *too small to sustain life.* The first may be regarded as the exact shape and figure of a short-lived female; and this may be contemplated as an equally true

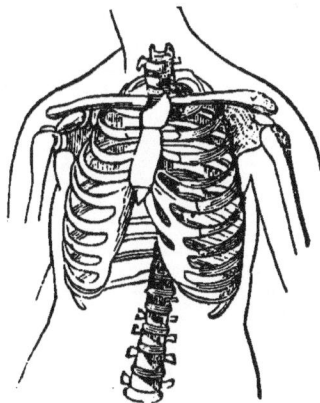

FIG. 17.

model of the frame of another, who, so far as life depends upon a well-formed body, would live to a good old age.

4. Round shoulders, narrow chests, small, weak lungs, and diseases of the spine are common results of bad habits of dress and posture. It may be said that *any* habit which tends to distort the frame-work of the body is so much direct injury to one or more of its organs and its functions.

QUESTIONS

FOR

EXAMINATION AND REVIEW.

QUESTIONS.

PART I

THE SKELETON.

Lesson I.

(a) — 1. What is the skeleton?

(b) — 1. Of how many bones is the skeleton composed? How does the number vary?

(c) — 1. State the use of the bones.

(d) — 1. Why are some bones long?

    2. Why are some bones short and thick?

    3. What bones are flat? irregular?

    4. What have you learned of their general form and adaptability?

    5. What is there peculiar about the structure of all the long bones?

Lesson II.

(a) — 1. Of what materials are the bones composed?

(b) — 1. Of what use is the mineral matter?

    2. Of what use is the animal matter?

(c) — 1. What is the nature of the substance of the bones?

    2. With what are the hollow portions filled?

    3. What of the bones in infancy? What change occurs later? What of the toughness of bone in early life? What of brittleness of bone in old age?

41

Lesson III.

(a) — **1.** Does bone once formed remain during life?

    2. At what points do bones grow?

    3. How are the bones supplied with nourishment?

(b) — **1.** What occurs as soon as a bone has been broken? What is the condition of the broken part after a few days? What time is required to complete the repair?

Rem. — What is said of the ease with which the change of bone may be noticed? What of mixing coloring-matter with the food of animals? How is extra strength given a broken bone?

Lesson IV.

(a) — **1.** At what points are the bones joined to each other?

    2. What is a joint?

(b) — **1.** Describe the covering of the ends of bones forming a joint.

    2. With what is the cartilage again covered? What oils the joints?

    3. By what are the bones at the joints held together?

(c) — **1.** Of what use are the joints?

Lesson V.

(a) — **1.** Mention the kinds of joints.

(b) — **1.** What motion do hinge-joints permit?

    2. Describe a ball-and-socket joint. What movements will it allow? Mention a joint of this kind.

Rem. — Tell what you know of the preparation of the fluid which oils the joints of animal bodies. What can you state about the durability of the joints?

Lesson VI.

Into how many and what classes are the bones of the skeleton divided?

(a) — 1. Into how many and what classes are the bones of the head divided? How many bones are in the skull? in the face? In each ear? How many teeth? Give the total number of bones in the head. *
(b) — 1. What do the bones of the skull form?
2. How are they united?
3. What is the form of the skull, and to what adapted?
4. What is said of the form and strength of the front of the skull? Why so formed?
5. What is the use of the packing between the bones?
6. What bones of the head are immovable?
(c) — 1. What protection do the bones of the skull afford?
2. Of what use are the bones of the ear?
3. Describe the joints and movement of the lower jaw.
4. Of what use are the teeth?

Lesson VII.

(a) — 1. Name the classes of bones of the trunk. How many bones in each of these classes? State the total number of bones in the trunk.
(b) — 1. How many and what cavities has the trunk?
2. What does the chest contain?
3. What does the abdomen contain?
4. By what are these cavities separated? By what enclosed?
Rem. — What is the diaphragm? Describe its form.

Lesson VIII.

(a) — 1. What is the thorax?
(b) — 1. What is the natural form of the chest?
2. By what are its walls formed?
3. What fills the spaces between the bones?
(c) — 1. With what instruments are the bones of the chest provided? For what purpose?
2. What organs are situated in the chest? What is their nature? and how protected?

Lesson IX.

(a) — **1.** Where is the spinal column situated? From what to what does it extend?

(b) — **1.** Of how many bones is the spinal column formed? What are these bones called, and why?

    2. What is found between each two bones? Of what use are these pads?

    3. Describe the form of the bones of the spine.

    4. How thick are they?

    5. Describe the spinal canal and its contents.

(c) — **1.** What does the spine support?

    2. Tell what is said of its variety of movement.

    3. What is said of the wonderful structure of the spine?

Rem. — What is said of the compression of the pads of the spine? Of the height of man at different times?

Lesson X.

(a) — **1.** What are the ribs, and how arranged?

(b) — **1.** To what are the ribs attached?

    2. What is said of the seven uppermost ribs?

    3. Describe the five lower ribs.

(c) — **1.** What is the use of the ribs?

    2. In what do they assist?

    3. What do the peculiar forms of the ribs give? What do the cartilages permit?

Lesson XI.

(a) — **1.** What is the pelvis, and where located?

(b) — **1.** Of what is the pelvis composed?

    2. Describe the form of the bones of the pelvis, and state what form they take. What bones are attached to the pelvis above and below?

(c) — **1.** What is the use of the pelvis?

    2. What sockets does it contain?

Rem. — What name is given to the hip-bones? What is said of the sacrum?

Lesson XII.

(a) Name the classes of bones in the upper extremities.
1. How many bones form the collar? What name is applied to them?
2. How many bones in the shoulder-blades? What are they also called?
3. How many bones in the upper arms? What called?
4. How many in the lower arms? What is each called?
5. How many bones in the wrist? Their name?
6. How many in the hands? What called?
7. How many in the fingers? Their name?

Lesson XIII.

(a) — 1. Locate the scapula. What familiarly called?
(b) — 1. What is the form, etc., of the scapula? In what is it embedded, and how kept in place?
2. Is it attached to the trunk?
3. To what is it connected, and where? Describe its socket.
(c) — 1. What is the use of the scapula?
2. In the formation of what does it aid? How?

Lesson XIV.

(a) — 1. Locate the clavicle.
(b) — 1. What is the shape of the clavicle? Why so named?
2. To what is it attached?
(c) What is the use of the clavicle?
Rem. — What is the effect of removing or breaking the clavicle? Have the lower animals collar-bones?

Lesson XV.

(a) — 1. Where, and at the junction of what bones, is the shoulder-joint?
(b) — 1. What bones form the shoulder-joint? What kind of joint is it?
2. Describe the construction of this joint.

(c) — **1.** What movements does this joint permit ?

Rem. — What is said of the dislocation of the shoulder-joint ? What if the socket were deeper ?

Lesson XVI.

(a) — **1.** How many bones in the upper arm ?　Give its name.
　　2. How many bones between the elbow and wrist ?　Give their names.
　　3. To what does the radius extend ?　The ulna ?
　　4. Describe the ulna, and give its position.　Locate the radius.
　　5. To what are the arms attached, and from what suspended ?
　　6. To what do the bones of the arm furnish attachment ?

Lesson XVII.

(a) — **1.** What bones articulate at the elbow ?　What kind of joint is formed there ?
(b) — **1.** Describe the union of the radius and ulna at the elbow.
　　2. What is the size of the ulna at the elbow ?　In what does it assist at that point ?
(c) — **1.** Describe the action of the bones at the elbow-joint.
　　2. How many and what movements does the elbow-joint permit ?

Lesson XVIII.

(a) — **1.** Where is the wrist located ?
(b) — **1.** Of what does the wrist consist ?　How are the bones arranged ?
　　2. With what does each of the rows of bones articulate ?
　　3. By what are these bones held together ?
(c) — **1.** What kind of joint is the wrist ?　What motions does it allow ?
　　2. What is said of the arrangement of its bone in regard to variety of movement ?　In regard to strength and elasticity ?

Lesson XIX.

(a) — 1. With what do the bones of the palm of the hand articulate? How many metacarpal bones in each hand?

2. With what bones do the other extremities of the bones of the palm articulate? What name is given the bones of the fingers? What bones, then, comprise the entire hand?

(b) — 1. With what bones do those of the palm articulate?

2. How are the first bones of the fingers joined to those of the palm? What two motions do these joints allow? What kind of joint do the other bones of the fingers form?

3. How are the first bones of the thumbs placed? What is said of their movements?

4. How many bones in a finger? In a thumb?

(c) — 1. What is said of the arrangement of the hand? To what is it adapted?

2. What do the numerous joints, etc., permit?

Rem. — To what being does the perfect hand belong? What does it confer upon man? Of what is the hand the principal organ?

Lesson XX.

(a) Write a table of the names and number of the bones of the lower extremities.

Lesson XXI.

(a) — 1. What bones articulate to form the hip-joint? What kind of joint is it?

2. Describe the structure of the hip-joint.

3. How tightly does the ball of the femur fit in its socket? *Note.* Describe the femur.

(b) — 1. What movements does the hip-joint permit?

(c) — 1. What bones articulate to form the knee-joint? What kind of joint is it?

2. Describe the patella, and tell how it is placed. Of what use is it?

3. How is the fibula placed ?   Of what use does it appear to be ?

(d) — 1. At what point does the knee permit bending of the leg ? What motions does it permit ?

Rem. — What is said of the displacement of the patella ?   Of its liability to receive blows ?

˙ Lesson XXII.

(a) — 1. What bones articulate to form the ankle-joint ?   What kind of joint is it ?

(b) — 1. What kind of movement does the ankle-joint permit ? What if this joint did not exist ?

(c) — 1. Of how many bones does the foot consist ?   Give the names of the bones.

2. To what is the foot similar in structure ?

3. What bones form the arch of the foot ?   Of what shape are they ?   How are they bound together ?   What does the arch allow ?

4. With what do the bones of the instep articulate ?   With what do the metatarsals again articulate ?

(d) — 1. What are the uses of the foot ?

2. What is said of the action of the foot ?   Describe its action.

Rem. — What is said of deformity of the foot being caused by shoes ?   What is the effect of high-heeled shoes ?

Lesson XXIII.

Write a classification of the bones of the skeleton, giving the number in each sub-class, in the following order: **(a)** Bones of the Head; **(b)** Bones of the Trunk; **(c)** Bones of the Upper Extremities; **(d)** Bones of the Lower Extremities.

Lesson XXIV.

(a) — 1. Upon what does the health of the bones depend ?

2. What effect has feebleness of health upon the bones of a child ?

3. Upon what do the size and strength of bones greatly depend?

(b) — 1. What tends to produce distortion of the spine? What organs occupy the right and the left sides of the cavities? In what condition is the stomach frequently? How does tight clothing curve the spine? How are the pads of the backbone affected?

2. What is said of the effect of tight clothing on many school-girls? What diseases are caused by tight clothing?

3. What effect has the constant bending forward of the head on the spine? What is said of bending over books in study, and of desks which are too high?

4. What are the common results of bad habits in dress and posture?

PART II.

DIGESTION

AND THE

MACHINERY OF DIGESTION.

"For now the cordial powers
Claim all the wandering spirits to a work
Of strong and subtle toil, and great event,
A work of time; and you may rue the day
You hurried with untimely exercise
A half-concocted chyle into the blood."

DIGESTION.

Lesson I.

(a) *Food and Hunger.* — **1.** Waste and worn-out material is constantly being cast out from our bodies. The lungs and the pores of the skin are busily engaged in this work.

2. If new material be not supplied to take the place of the worn-out substances, the body would dwindle and die. Without food, a man will starve in a few days.[1]

3. When the body needs material to take the place of that which is worn out, the nerves of the stomach become active in a peculiar way; and, when the sensation is carried to the brain, we recognize it as *hunger*.

---

[1] Dr. Tanner's experiments prove, that, under favorable circumstances, a strong man may live for many days without food. Dr. Tanner succeeded in abstaining from food forty days, but was fully supplied with fresh water and air, and with pleasant company, which aided very much in stimulating him in his long fast.

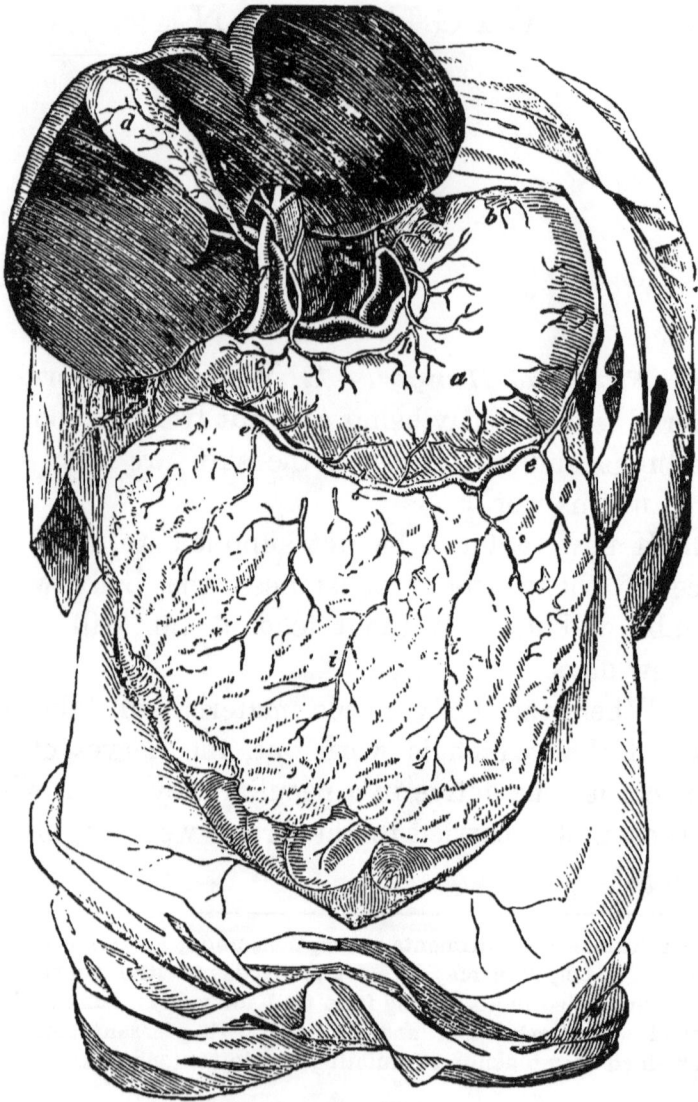

FIG. 18.

EXPLANATION OF FIG. 18.

In this view of the abdomen, *d* is the gall-bladder, lying on the under side of the liver, the dark mass to which it is attached.

h is the *coronary* artery, which supplies the stomach, *a, b, c,* with blood. The curve of the stomach is well shown.

e, e, the arteries which supply the caul, marked *i, i,* which falls down from the front of the stomach, over the intestines, like an apron.

g, a vessel of the liver. The *pancreas* is behind the stomach.

54

(b) *Food and Force.*—**1.** All the strength of our bodies comes from the food we eat. After the food has gone through the different processes of digestion, it gives up to the blood properties that supply the body with nourishment and strength. Just as new fuel feeds the fire, so does food keep up the forces of the body.

2. The waste of bodily substance differs in different persons and under different circumstances. Great bodily action causes great waste or wearing-out of the particles.

Lesson II.

FOOD. — *Concluded.*

(a) *Why Food must be Digested.*—**1.** Food is not in condition to be taken into the blood from the stomach as soon as it reaches that organ.

2. The food must be changed in various ways to prepare it for the use of the body. These changes are called *digestion*.

(b) *The Digestive Machinery or Organs.*—**1.** The organs of digestion are the *mouth, teeth, tongue,* and *lips.*

2. The *salivary* glands.

3. The *pharynx.*

4. The *œsophagus.*

5. The *stomach.*

6. The *lacteals.*

7. The *duodenum.*

Remarks. — Besides these organs, there are some others that render assistance in the work of digestion. These are the liver and the pancreas, which supply fluids that aid in preparing the food to become blood. The liver supplies a bitter fluid called *bile,* and the pancreas supplies the *pancreatic* juice.

Lesson III.

THE MOUTH.

(a) *Work of the Mouth.* — **1.** The lips and cheeks form the outward walls of the mouth. They retain the food when it is put into the mouth.

2. The teeth cut and grind the food to a fineness suitable to the stomach.

3. The tongue rolls the food about, and keeps it in its place between the teeth.

4. The salivary glands (*sacks*) excrete (*give out*) saliva or spittle to moisten the food and aid in bringing out its taste.

Remarks. — Nature has provided an organ to supply the proper kind and quantity of liquid to moisten our food in the mouth; and it is therefore unnecessary, if not injurious, to deprive her of her office by taking a " swallow " of tea, coffee, or water, with every morsel of food. Drink should be taken after eating, or, better still, before eating.

Lesson IV.

THE SALIVARY GLANDS.

(a) *Position of the Salivary Glands.*—1. In the cheeks.

2. Under the tongue.

3. Under the jaw.

(b) *Construction of the Salivary Glands.*—1. They are small sacks.

2. They open into the mouth through very small tubes.

(c) *Work of the Salivary Glands.*—1. They pour out saliva (*spittle*) whenever the tongue and cheeks are put in motion.

2. When the tongue and cheeks are not in motion, they let out no more saliva than enough to keep the mouth moist.

3. The presence of any thing in the mouth, any motion of the jaws, the chewing of our food, tobacco, etc., excites these glands and causes a flow of saliva.

4. The office of these glands is to moisten food and to keep the mouth moist.

Remarks.—All motions of the tongue, cheeks, and jaws are usually needless, except when we eat, drink, or talk. These organs are under our control, and the flow of saliva is under our command. The chewing of tobacco, gum, etc., keeps the glands unduly excited; and from being over-worked, they may become unable to properly perform their duty in moistening the food.

The moistening of the food by these glands is called *insalivation.* The grinding of the food by the teeth is called *mastication.*

Lesson V.

THE PHARYNX.

(a) *Position.* — 1. The *pharynx* is located back of the mouth, and back of the palate.

2. It connects the mouth with the œsophagus.

(b) *Construction.* — 1. It spreads out like a funnel behind the palate, and it is open to receive the food.

2. It has elastic walls formed of muscles.

(c) *Work.* — 1. The office of the pharynx is to receive the food from the mouth, and to aid in swallowing.

2. It acts as a funnel to the œsophagus.

Lesson VI.

THE ŒSOPHAGUS.

(a) *Position.* — 1. The *œsophagus* extends from the pharynx downward to the stomach.

2. It is located between the trachea (*windpipe*) and the spinal column.

3. Its lower extremity opens into the stomach.

(b) *Construction.* — **1.** It is a soft tube about nine inches long, and rather less than an inch in diameter.

2. It is covered with two layers of muscles, one of which runs lengthwise, and the other winds around it successively from top to bottom.

3. These muscles have a power of contraction, or of drawing themselves up like the earthworm, and of relaxing themselves, and being stretched out loosely.

(c) *Work.* — **1.** When the food is thrust backward by the tongue, it passes into the pharynx, which closes upon it and forces it downward into the œsophagus.

2. The uppermost ring of muscle contracts and closes the upper end of the œsophagus, thus preventing a return of the food upward.

3. The next band of muscle contracts and forces the food downward; then the third band does the same, and each successive one continues the work till the food is forced downward into the stomach.

4. While one band is contracting, the next one below is relaxing to admit the food.

Remarks. — Vomiting is performed in the same way, except that the order is reversed. The lowest band contracts first, and then the next above, thus forcing the contents of the stomach upwards to the mouth.

Lesson VII.

THE STOMACH.

(a) *Position.* — 1. The *stomach* is placed on the left side of the abdomen, just below and within the lower ribs.

(b) *Construction.* — 1. The stomach is a long, round, and somewhat irregularly-shaped sack. Its shape is like that of a bagpipe or shot-pouch.

2. It has two openings, — one towards its left extremity, where the œsophagus opens into it, and the other at the right extremity, where it opens into the duodenum, the upper portion of the alimentary canal.

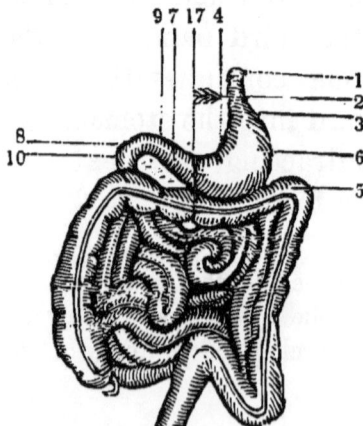

FIG. 19.

EXPLANATION OF FIG. 19.

1, the œsophagus.
2, the left opening of the diaphragm.
3, the cardiac orifice of the stomach.
4, the small curvature of the stomach.
5, the great curvature of the stomach.
6, the fundus of the stomach.
7, the pyloric orifice.
8, 9, 10, the duodenum, divided into three portions.

3. The stomach is fleshy, and very soft and flexible.

4. It is composed of three coats, or layers; viz.,

the outer or *peritoneal* coat, the middle or *muscular* coat, and the inner or *mucous* coat.

5. The outer or peritoneal coat is very tough and strong, and, being attached to the backbone and sides of the abdomen, it holds the stomach in place.

6. The middle or muscular coat is composed of muscles, some extending lengthwise, and others circularly. These muscles have a power of contraction and expansion.

7. The inner or mucous coat is loose, soft, and spongy, and covers the inner surface of the stomach. It is not elastic. When the stomach is full, this coat is smooth; but when the stomach is nearly empty, it is drawn into folds or furrows.

Remarks. — The appearance and structure of the different coats of the stomach may be studied by examining the stomach of an animal, an ox or a cow, when prepared for food, and called *tripe.*

Lesson VIII.

THE STOMACH. — Concluded.

(c) **Work.** — **1.** The *gastric juice* is prepared within the walls of the stomach, and thrown out from the inner or mucous coat.

2. The gastric juice dissolves certain parts of the food, and helps to prepare it to be absorbed into the blood. This juice pours into the stomach in con-

siderable quantities when food enters it, and causes
a fermentation which changes the solid food into a
liquid. This process is called *chymification;* and the
pulp into which the food is thus changed is called
chyme.

3. During digestion in the stomach, the muscular
coat contracts and expands in order to contract
again, thus keeping the contents of the stomach in
constant motion, mixed with the gastric juice, and
moved toward the *pylorus.*

4. The chyme is now ready to pass from the
stomach to its second stage of digestion in the *duodenum.*

Remarks. — The process of digestion requires the natural
heat of the body. It has been found by tests that the temperature of the stomach is about 100°.

If cold liquids be swallowed, the temperature will be lowered,
and digestion will be stopped until the temperature again rises
to the proper height.

Different kinds of food require different lengths of time for
digestion.

About 1822 a young soldier named Alexis St. Martin, employed in the service of the United States, was badly wounded
by the bursting of a gun, which tore away the flesh of the abdomen and a part of his stomach. He got well; but the hole
in his stomach did not heal, and, by pulling aside a piece of
skin, one could look into the stomach, and see its action. Dr.
Beaumont tried many experiments with him, and, by putting
different kinds of food into St. Martin's stomach through the
hole left by the wound, found out how long different kinds of
food required for digestion. A thermometer was passed into
the stomach, and the temperature ascertained. Dr. Beaumont

thus had an opportunity for observation and experiment which probably no other man has ever had; and to these experiments we owe most of our knowledge of digestion.

Table giving the Length of Time required for the Digestion of a Few of the most Ordinary Kinds of Food.

ARTICLES.	CONDITION.	TIME.	
		Hours.	Minutes.
Pork, fat and lean	Roasted	5	15
Suet, beef, fresh	Boiled	5	30
Cabbage, with vinegar . .	Boiled	4	30
Ducks, domestic	Roasted	4	00
Ducks, wild	Roasted	4	30
Cheese, old, strong . . • .	Raw.	3	30
Eggs, fresh	Boiled hard . . .	3	30
Eggs, fresh	Raw.	2	00
Chicken, full grown . . .	Fricaseed	2	45
Bread, white	Baked	3	30
Potatoes, Irish	Boiled	3	30
Codfish, dry	Boiled	2	00
Soup, bean	Boiled	3	00
Soup, barley	Boiled	1	30
Rice	Boiled	1	00
Oysters, fresh	Raw.	2	55
Apples, sweet	Raw.	1	30
Dumpling, apple	Boiled	3	00

Lesson IX.

THE INTESTINAL CANAL. — THE PYLORUS.

(a) *The Intestinal Canal, and Position of the Pylorus.* — **1.** The *intestinal canal* is a continuation of the stomach, and consists of the large and small intestines.

2. The *pylorus* is an opening, or valve, situated at the right or smaller end of the stomach.

3. *Pylorus* signifies *doorkeeper*, and is so named because it allows the contents of the stomach, which have been properly prepared, to pass out, while other portions, not prepared, are held back.

(b) *Construction of the Pylorus.* — **1.** The pylorus is a muscular valve, consisting of a band or ring of muscle, which surrounds the opening at the right end of the stomach.

(c) *Work of the Pylorus.* — **1.** While the stomach is engaged in its work, the pylorus draws itself firmly around the opening, like a shir-string, and prevents undigested food from passing out.

2. As soon as the stomach has performed its work upon any portion of the food and reduced it to chyme, it carries the chyme to the opening where the pylorus is placed. The pylorus loosens its hold and permits the chyme to pass out freely; but, when undigested portions of the food present themselves, the pylorus contracts, closes the opening, and prevents them from leaving the stomach.[1]

3. When the stomach is supplied with food that it cannot digest, it endeavors to free itself of its stubborn tenant by forcing him out through the pyloric opening; but the pylorus resists, and sturdily refuses to allow the door to be opened. After repeated endeavors of the stomach to digest the food, and persistent refusals of the pylorus to permit it to pass out undigested, this faithful servant becomes weakened by the struggle. The undigested food is thrust through the pylorus and passes into the alimentary canal, causing irritation and discomfort on its way.

4. These struggles with indigestible food finally result in great weakness of the machinery of digestion and cause dyspepsia, or other disorders of the stomach.

Lesson X.

THE DUODENUM.

(a) *Position.*—1. The *duodenum* is situated just beyond the pylorus, at the upper end of the alimentary canal.

---

[1] In a recent case, the pylorus closed upon a prune-stone that had been accidentally swallowed, and held it so tightly in its folds as to cause inflammation and death.

2. The pylorus opens into the duodenum from the stomach.

(b) *Construction.* — 1. The duodenum is so named because it measures nearly twelve finger-breadths in length. It is bent upon itself, and fastened against the back wall of the abdomen.

2. It is composed of three coats similar to those of the stomach.

(c) *Work.* — 1. When the chyme enters the duodenum, two juices, the *bile* and the *pancreatic* juice are poured into the duodenum and mingle with the chyme, just as the gastric juice mingles with the food in the stomach.

2. These juices aid in liquefying the chyme, and change it into a milky fluid called *chyle*. This process is called *chylification*.

3. The inner lining of the duodenum, the *mucous membrane*, gives out (*excretes*) a slimy fluid which moistens the inner surface of the duodenum and protects it from any irritating quality of the contents.

Remarks. — The moment chyle is formed, digestion proper may be considered as completed, though the chyle must still be absorbed, in its course through the smaller intestine, before it mingles with the blood.

Lesson XI.

THE LACTEALS.

(a) *Position.* — 1. The inner lining (*mucous membrane*) of the intestinal canal is filled with myriads of pores, or openings into hair-like tubes, which run outward through the walls of the intestine.

2. These little tubes are called *lacteals* (*lactis*, milk).

(b) *Construction.* — 1. These little tubes, when they first start from the inside of the intestine, are extremely small, but afterward they unite and become larger and fewer.

2. The larger tubes again unite and form other and still larger ones, until all unite in one large tube named the *thoracic lacteal duct.*

3. The mouths of the lacteals are so small as to be invisible, except by aid of a powerful microscope.

(c) *Work.* — 1. The office of the lacteals is to absorb, or suck up, some of the nourishing portions of the chyle.

2. They convey the nourishing portions of the chyle into the *thoracic duct.*

3. These tubes, large and small, and the thoracic duct form what is called the *lacteal system.*

Lesson XII.

THE THORACIC LACTEAL DUCT.

(a) *Position.* — 1. This duct extends from the abdomen to the upper part of the chest, along the inner side of the backbone, and, bending forward, opens into the great vein at the right side of the heart.

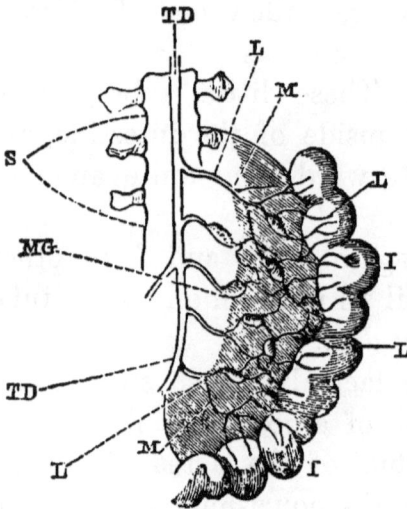

FIG. 20.

EXPLANATION OF FIG. 20.

A portion of the *thoracic duct*, marked T D above, and T D below, lying in front of and in contact with the spine, S.

By the side of I I is seen a portion of intestine attached to the mesentery, a kind of membranous ruffle, around the border of which the entire tube of the intestine is fastened.

L L show a lacteal vessel running from the inside of the intestine, charged with a milky fluid which is conducted into the mesenteric glands, seen lying between the two folds of that membrane. In these the chyle is essentially changed in character, and perhaps receives additional fluid from the gland itself. From these the fluid next passes on through the excretory ducts, M M, which join the main trunk of the thoracic duct.

(b) *Construction.* — 1. It is a tube about as large as a goose-quill.

2. It is formed by the union of the lacteals.

(c) *Work.* — 1. Its office is to convey the nourishing portions of the chyle from the digestive apparatus to the blood-vessels. It pours the chyle into the vein leading to the heart.

Remarks. —The work of digestion consists of three distinct parts; viz., *mastication and insalivation* in the mouth, *change of food into chyme* in the stomach, and the *change into chyle*, and separation of the worthless parts, in the duodenum. How these fluids exercise their latent powers in giving force and strength is precisely known only by the Creator. Enough, however, is known by us to guide us in the selection of proper food and in the proper use of the organs of digestion.

Lesson XIII.

HINTS ABOUT EATING.

1. The stomach does its work best when the mind is at ease and the body is rested. Children often rush to their meals when heated and excited by play, and, though it does not at the time appear to injure them to take food while in this condition, still they will almost certainly suffer for it in time. It is much better to give the body time to rest and become cool, and the nerves a chance to become quiet, before eating. The food will not only taste better, but will also digest better.

2. *Meals should always be eaten at regular hours.* Great injury is often done to the health by the habit of eating irregularly and between meals. By this practice the stomach is kept at work almost con-

stantly. The stomach needs time for rest as well
as the other parts of the body do, and, if it be all
the time worried with extra work, it must and will
become tired out and *worn out*, and consequently
unable to do its work. A very large part of the
sicknesses of the body is caused by abuse of the
stomach. About five hours should elapse between
meals, and meals should be taken at the same hours
each day.

3. *We should not eat hurriedly.* The food should
be properly masticated, and there is no worse habit
connected with digestion than that of swallowing
our food in haste. The few minutes gained by this
habit are sure to be dearly paid for by and by.

4. When food is taken into the stomach, the blood
rushes toward that organ and raises its heat. This
being so, we should keep the body as quiet as possible for a time after eating, for violent exercise
always causes a rush of blood to the surface of the
body; and, as this draws the blood away from the
stomach, it does not have the heat required to digest
the food properly. It is not necessary that we remain perfectly still, for moderate exercise which does
not call the blood away from the stomach will harm
no one. Such play as running, leaping, jumping
rope, etc., should not be indulged in for at least a
half-hour after eating.

5. Brain-work, also, causes the blood to flow toward the head; and children should not engage in
hard study for at least an hour after a hearty meal.

QUESTIONS

EXAMINATION AND REVIEW.

c

QUESTIONS.

DIGESTION.

Lesson I.

(a) — 1. What are our bodies constantly giving off ?
2. Why is new material necessary? What would result if we were deprived of food ?
3. Describe what takes place in the stomach when the body needs new material.

(b) — 1. What supplies all our strength? What does it supply, and when? To what may this process be compared?
2. What is said of the waste of bodily substance in different persons, etc. ? What causes extra waste of substance? *Note.* Relate what is said of Dr. Tanner's experiment.

Lesson II.

(a) — 1. What is said of the condition of food when it first enters the stomach ?
2. What must happen to the food ? What are these changes called ?

(b) — Name the organs of digestion.

Rem. — What is said of certain other organs ? Of the pancreas and liver ?

Lesson III.

(a) — **1.** What do the lips and cheeks form, and what work do they perform in digestion?
2. State the work of the teeth.
3. State the work of the tongue.
4. State the work of the salivary glands. In what organ do these processes occur?

Rem. — What has nature supplied to moisten our food? What is said of taking a "swallow" of liquid with our food? When should drinking be done?

Lesson IV.

(a) — **1, 2, 3.** Locate the salivary glands.
(b) — **1.** What are the salivary glands?
2. Into what and by what do they open?
(c) — **1.** What do they pour out, and when?
2. What of their action when the tongue and cheeks are at rest?
3. What things tend to excite these glands to action?
4. What is the office of the salivary glands?

Rem. — What is said of certain motions of the mouth, etc., when we are neither eating nor talking? What is insalivation? What is mastication?

Lesson V.

(a) — **1.** Locate the pharynx.
2. What two organs does it connect?
(b) — **1.** What is the form of the pharynx?
2. Of what are its walls formed?
(c) — **1.** What is the office of the pharynx?
2. What relation does it bear to the œsophagus?

Lesson VI.

(a) — **1.** How is the œsophagus situated?
2. What is its position in regard to the trachea and the spinal column?

3. Into what does its lower extremity open ?

(b) — **1.** What is the form of the œsophagus ? What is its size ?

2. With what is the œsophagus covered ? How are these layers of muscles arranged ?

3. Describe the action of these muscles.

(c) — **1.** State how food reaches the œsophagus.

2. Describe the first act of the œsophagus in swallowing food, etc.

3. Describe the successive acts of the œsophagus in swallowing.

4. What occurs while one band of muscle contracts ? Why is this ?

Rem. — What is remarked of the process of vomiting ?

Lesson VII.

(a) — **1.** Give the position of the stomach.

(b) — **1.** What is the stomach ? What does it resemble in shape ?

2. State the number of its openings, and locate each.

3. What is the nature of the substance of the stomach ?

4. Of what is the stomach composed ? Name the coats in their order.

5. Describe the outer coat, stating its attachments.

6. Describe the nature of the middle coat.

7. Describe the inner coat.

Rem. — How may the appearance of the stomach and its structure be readily studied ?

Lesson VIII.

(c) — **1.** Where is the gastric juice prepared ? From which coat is it thrown out ?

2. Of what use is the gastric juice ? What effect has it upon the food ? What is this process called ? What name is given to the pulpy mass ?

3. Describe the action of the muscular coat of the stomach during digestion.

4. For what is the chyme now ready, and into what does it pass ?

Rem. — What is remarked of the temperature of the stomach? What of the effect of swallowing cold liquids during digestion? What of the length of time required for the digestion of different kinds of food? Can you give the length of time required for any particular kinds of food? Relate the account given of Alexis St. Martin.

Lesson IX.

(a) — 1. What is the intestinal canal?
   2. What is the pylorus?
   3. What does pylorus signify, and why is it so named?
(b) — 1. Describe the construction of the pylorus.
(c) — 1. Describe the work of the pylorus while the food is being changed to chyme.
   2. What becomes of the food as soon as it is changed into chyme? What does the pylorus then?
   3. If undigested food presents itself, what does the pylorus?
   4. In what do the struggles with indigestible food result?
   Note. Relate the action of the pylorus in closing upon a prune-stone.

Lesson X.

(a) — 1. Locate the duodenum.
   2. What opens into it from the stomach?
(b) — 1. Why is the duodenum so named? How is it bent, and how attached?
   2. Of what is it composed?
(c) — 1. What occurs when the chyme enters the duodenum? Give the names of the juices.
   2. What effect have these juices on the chyme? What is this process called?
   3. Describe the work of the inner lining of the duodenum.
Rem. — When is digestion completed? What still remains to be done?

Lesson XI.

(a) — 1. With what is the inner lining of the intestinal canal filled?
   2. What are these little tubes called?

(b) — **1.** What is the size of the lacteals at first? Afterward?

2. What do the largest branches of the lacteals finally form?

3. What do you know of the size of the mouths of the lacteals?

(c) — **1.** What is the office of the lacteals?

2. Whither do they convey the nourishing particles?

3. What do the lacteals and the thoracic duct together form?

Lesson XII.

(a) — **1.** From what to what does the thoracic duct extend? Along what does it extend? Into what does it finally open?

(b) — **1.** Describe the thoracic lacteal duct.

2. By the union of what is it formed?

(c) — **1.** What is the office of this duct? Into what does it pour the chyle?

Rem. — Briefly state the processes of digestion in their order. What is remarked of the way in which the nutritive liquid obtains its power?

Lesson XIII.

1. When does the stomach perform its work best? What is said of the practice of children in rushing to their meals when heated and excited by play?

2. What is said of regular hours for meals? Why should food be taken at regular intervals? What is said of rest for the stomach? From what does a large part of the sickness of the body arise? About how many hours should elapse between meals?

3. What is said of hurried eating?

4. What is said of the rush of blood toward the stomach? Why should we remain quiet for a time after eating? What should not be engaged in for at least half an hour after eating?

5. What effect has brain-work on the flow of blood? What is said of study after eating?

PART III.

THE BLOOD

AND ITS

CIRCULATION.

" The blood,—the fountain whence the spirits flow,
The generous stream that waters every part,
And motion, vigor, and warm life conveys
To every particle that moves and lives."

THE BLOOD.

Lesson I.

(a) *What and Where.* — 1. The *blood* is the liquid which circulates through the different parts of the body, and conveys to them materials for their nourishment.

2. It is found in every part of the body except the outer skin (*cuticle*), the nails, the hair, the cornea of the eye, etc.

3. The average quantity in the human body is about eighteen pounds.

(b) *Composition.* — 1. The blood consists of a colorless liquid, called the *plasma*, in which float countless little circular bodies or disks. These little bodies are the *corpuscles* (*i.e.*, little bodies) of the blood, and they are of a bright-red color. In human blood these corpuscles are $\frac{1}{3200}$ of an inch in diameter, and about $\frac{1}{12800}$ of an inch thick.[1]

---

[1] The size of the corpuscles in human blood is not the same as that of the corpuscles in the blood of lower animals. This fact has aided greatly, in trials for murder, in discovering whether bloodstains on clothing and weapons were made by human blood.

2. The corpuscles are a little heavier than the plasma; and, when the blood is drawn from a vein, they sink toward the bottom. Examination of blood that has been drawn a little while will discover the colorless plasma, in which but few corpuscles remain on the top. The corpuscles tend to collect like rolls of coin.

3. The plasma, or *nutritive fluid*, is composed of water richly laden with materials derived from the food.

Lesson II.

USES OF THE BLOOD.

1. The *blood* contains the materials for making every organ of the body.

2. The *plasma* contains mineral matter to supply the bones, and also animal matter to deposit with the muscles.

3. The *corpuscles* contain the oxygen and certain other materials necessary to the life of the body. They are the air-cells of the blood.

4. In short, the blood both carries new materials to all the organs, and removes worn-out particles of matter. It conveys *oxygen*, and removes *carbonic acid gas*.

5. The blood is in constant motion during life. From the heart as a centre, a current is always flow-ing toward the different organs, and from these organs a current is constantly returning to the heart. This movement is called *the circulation of the blood.*[1]

6. The organs of the circulation are,

The *heart,*

The *arteries,*

The *veins,*

The *capillaries.*

Remarks. — If blood from a living animal be injected into the veins of one that is very weak from loss of blood, strength and new life return to the seemingly lifeless animal. This operation, called *transfusion,* has been practised upon man with similar results. It is still practised in cases where there has been a great loss of blood.

Lesson III.

THE HEART.

(a) *Position.* — **1.** The *heart* is the organ which propels the blood, and is situated just to the left of the centre of the chest.

(b) *Construction.* — **1.** The heart is a hollow, muscular organ, shaped like a strawberry, and suspended with the point downwards. Its size is roughly estimated to be equal to that of the fist.

---

[1] The circulation of the blood was discovered by William Harvey in 1619.

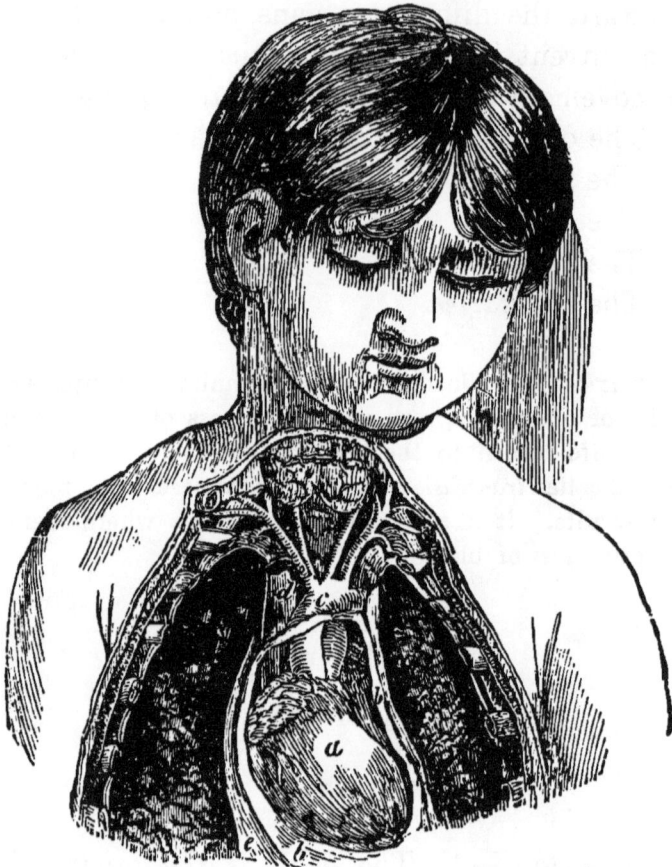

FIG. 21.

EXPLANATION OF FIG. 21.

a, the *heart* in its natural position, the sternum being taken away, and the pericardium laid open in front.

c is the arch of the *aorta*, the artery from which all others arise.

b, b, i, the *pericardium*.

d, the *descending cava*, or great vein, that returns the blood from the head and arms into the right auricle.

2. It is surrounded by a loose sack of membrane, the *pericardium* (*peri*, about, *cardia*, the heart). The pericardium is as smooth as satin, and gives out a liquid which keeps it moist and pliable.

3. The heart is partitioned into four chambers. The two upper ones are called *auricles* (*aures*, ears), because of the shape of the flaps on their outside walls. The lower chambers are called *ventricles* (*ventriculus*, a cavity).

4. The auricle and the ventricle on the same side communicate with each other by means of openings (*valves*); but the right and left sides of the heart are entirely separated by a muscular partition in which there is no opening.

5. The walls of the ventricles are thicker than those of the auricles. This is a wise provision; for it is by the powerful action of the ventricles that the blood is forced to the remotest regions of the body.

6. The auricles need much less power, for they simply discharge their contents into the ventricles, which are near at hand, and their walls are not so thick.

7. The valve between the right auricle and the right ventricle consists of three flaps of muscle, and is called the *tricuspid* valve (*tri*, three, *cuspides*, points). The valve between the left auricle and left ventricle consists of two flaps, and is called the *bi-cuspid* valve. The passages from the ventricles into the arteries are closed by half-moon-shaped valves, called *semi-lunar* valves.

Lesson IV.

THE HEART. — *Concluded.*

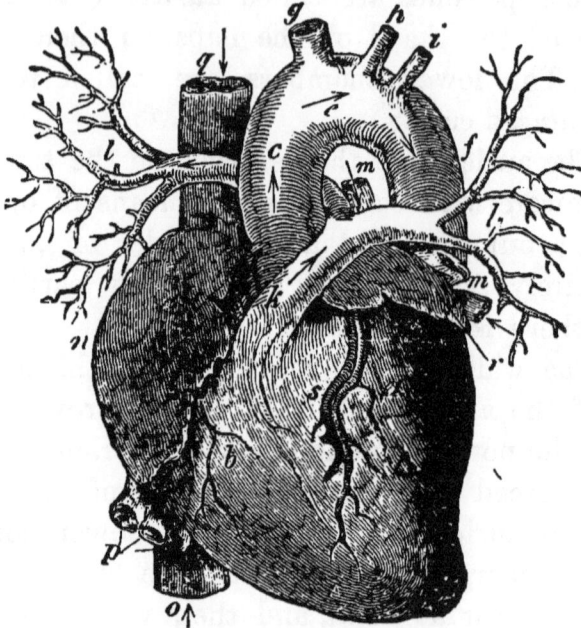

Fig. 22.

EXPLANATION OF FIG. 22.

The double heart of man.

q, the *descending vena cava.*

o, the *ascending vena cava.*

n, the *right auricle.*

b, the *right ventricle.*

k, the *pulmonary artery.*

. *l, l*, the *right* and *left* branches of this artery, going to the lungs on either side of the chest.

m, m, the *veins* of the *lungs*, which

return what the artery sent in, to *r*, the *left auricle.*

a, the *left ventricle.*

c, e, f, the *aorta*, or great artery of the body, rising out of the left heart.

g, the *arteria innominata.*

h, the *subclavian artery*, going to the left arm.

i, the *carotid artery*, which goes up the side of the neck to the head.

Note. — The arrows show the course the blood moves in each of the vessels demonstrated with the heart: *n*, the right auricle; *m, m*, veins of the lungs; *s*, the left coronary artery; *p*, the veins returning blood from the liver and bowels.

(c) *Work.* — **1.** The action of the heart consists of alternate contractions and dilations. During contraction, the walls come forcibly together, and thus the blood is driven out. In dilation or expansion, the walls open or separate, and thus make room for a new supply of blood.

2. The contraction of the right *auricle* drives the blood into the right *ventricle :* the right *ventricle* then contracts, and forces the blood through the *pulmonary artery* into the lungs.

3. Leaving the lungs where the blood is purified, it returns by four *pulmonary veins* to the left *auricle ;* the contraction of the left *auricle* drives the blood into the left *ventricle ;* the left *ventricle* contracts, and drives the blood into a large artery called the *aorta,* the branches of which convey it to all parts of the body, except the lungs, to which it is returned, as first described, after circulating throughout the entire body.

Remarks. — The heart itself is supplied with blood for its nourishment by two arteries which spring from the root of the aorta.

What is known as the *beating* of the heart is caused by the striking of the apex (lower end) of the heart against the pericardium, or sack which encloses it, and through it against the walls of the chest.

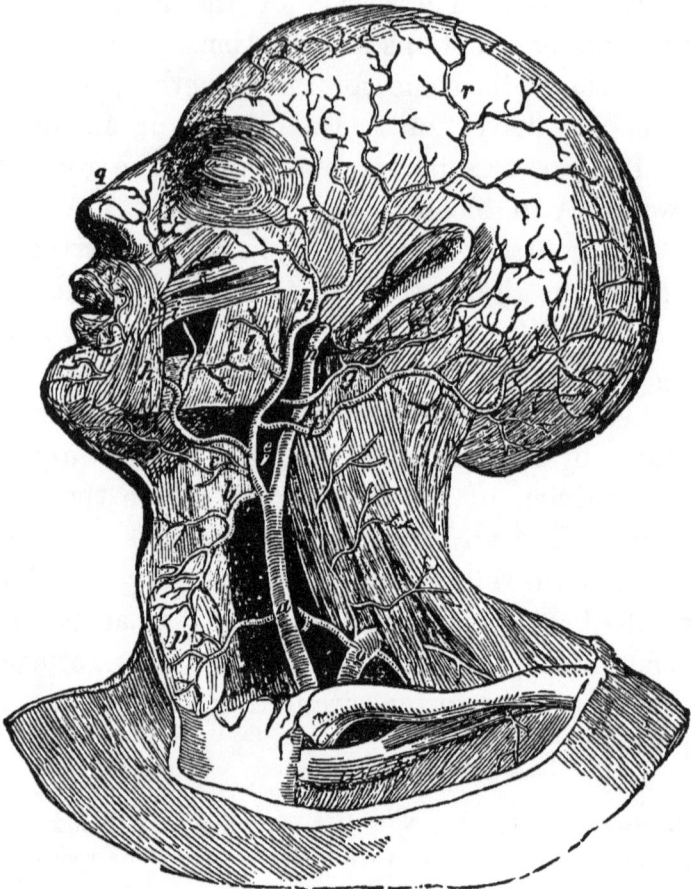

FIG. 23.

EXPLANATION OF FIG. 23.

a, the trunk of the *carotid artery*, which when compressed causes apoplexy and death.

f, the *occipital artery*, going to the muscles of the back of the head.

b, the *larynx*, or vocal box.

n, the *external carotid*, branching outward.

k, the *temporal artery*, felt beating in the temple.

q, the *nasal artery*.

r, the termination of the *temporal artery* in twigs on the top of the head.

Lesson V.

THE ARTERIES.

(a) *Position.*—1. The *arteries* are tubes which spring from the heart. The branches of one great artery extend throughout the body, while another and its branches extend to the lungs.

2. The large arteries and their principal branches are generally situated far beneath the surface, and their location gives them security from all ordinary danger. Many of them are found close to the bones, or running through safe passage-ways. The skin, hair, teeth, and bones are all provided with arteries.

(b) *Construction.*—1. The general arrangement of the arteries resembles that of a tree; the great artery being the trunk, and its divisions, the limbs and twigs, continually growing smaller.

2. Arteries are composed of three coats. The *internal coat* is smooth, polished, and moistened with an oily fluid to aid the easy flow of the blood. The *middle coat* consists of circular fibres, which are yellowish, dry, and elastic. This coat contracts and expands. The *outer coat* is dense and very elastic: it is so closely connected with the middle coat as to be difficult of separation from it. The coats are nourished by many capillaries.

(c) *Work.*—1. The arteries, being elastic, expand and contract at every beat of the heart, thus aiding to keep the blood in regular and constant motion on its way through them. They carry the pure, nourishing blood to all the organs of the body.

Remarks. — The flow of blood from an artery when cut differs from that of a vein. When an artery is cut, the blood spurts out "by jerks," at every impulse given it by the heart, while the flow from a vein is slow and regular.

Lesson. VI.

THE CAPILLARIES.

(a) *Position.* — 1. The *capillaries* (*capilla*, a hair) are situated between the ends of the arteries and the beginnings of the veins.

(b) *Construction.* — 1. Capillaries are so much a part of both arteries and veins that it is impossible to tell where an artery ends, or where a vein begins. They are constructed like a fine net whose meshes are composed of tubes so small that they cannot be seen with the naked eye. They are only about $\frac{1}{3000}$ of an inch in diameter.

2. They are placed so close together that the entire flesh is filled with them, and the prick of the finest needle would break great numbers of them.

(c) *Work.* — 1. The capillaries take up the nourishing particles of the arterial blood, and convey them to the flesh, muscles, bones, and to every part of the body in which they are located.

2. They receive the oxygen from the corpuscles of the blood, and in return give up carbonic acid gas and other impure and waste matter to the veins, which spring out of the capillaries. Thus they perform a double work in serving both arteries and veins.

————

Lesson VII.

THE VEINS.

(a) *Position.*—**1.** The *veins* begin in the capillaries. Some are located deeply, and accompany the arteries: others are situated just under the skin, and do not follow the direction of the arteries. They are found in every organ of the body.

(b) *Construction.*—**1.** The walls of the veins are not so thick as those of the arteries. They are formed of different coats; the inner one being, like that of the arteries, smooth and polished.

2. The inner coats are provided with folds, which, when extended, partly close the tube. These folds or *valves* are so arranged as to permit the blood to pass freely toward the heart, but, by letting themselves down in the passage-way, prevent its backward flow.

3. The veins are very small at first; but, as they leave the capillaries, they unite, and increase in size while they diminish in number. At last they all become united in one great vein,[1] which divides, just before reaching the heart, into two great branches, one of which is named the *vena cava ascending*, and the other the *vena cava descending*. These branches open into the right auricle.

(c) *Work.* — 1. The veins may be called the sewers of the blood. They conduct the impure blood from the capillaries, and carry it back to the heart, from which it was first sent out on its mission, *thus completing the circulation.*

Remarks. — The rapidity with which the blood moves through the blood-vessels is influenced so much by our emotions and by circumstances, that it cannot be exactly determined. It has been estimated that the whole volume of blood passes through the heart in about two minutes.

Lesson VIII.

RECAPITULATION. — MOVEMENTS OF THE BLOOD.

1. From the right *auricle* to the right *ventricle*.

2. From the right *ventricle*, through the *pulmonary artery*, into the *lungs*.

3. From the *lungs*, through the *pulmonary* veins, into the left *auricle*.

4. From the left *auricle* into the left *ventricle*.

----

[1] The great vein receives the new nourishing matter from the thoracic duct, and pours it, together with the old blood, into the heart.

5. From the left *ventricle* into the *aorta* and arteries.

6. From the *arteries* into the *capillaries.*

7. From the *capillaries* into the *veins.*

8. From the *veins* back again into the right *auricle* of the heart, *thus completing the circulation.*

Lesson IX.

WHAT HASTENS THE CIRCULATION.

1. In an adult in a normal condition, the heart beats about sixty times a minute, and consequently the pulse an equal number of times.

2. *The heart beats more rapidly during digestion,* and *under the influence of alcohol, coffee,* or other excitants.

3. *Mental labor also quickens the action of the heart.*

4. *Muscular exercise and violent efforts quicken the action of the heart,* and increase the rapidity of the circulation.

5. *The action of the heart is greatly hastened by fever.*

6. Causes which tend to unduly excite and prolong increased action of the heart should be avoided ; for the heart, like any other muscle, may be overtasked and weakened, and thus rendered incapable of performing its ordinary work.

Lesson X.

WHAT RETARDS THE CIRCULATION.

1. During sleep the action of the heart is less rapid, and it shares somewhat in the repose of the other organs.

2. *Abstaining from exciting mental labor, muscular effort, and from the use of alcohol, coffee, and other stimulants,* retards undue activity of the heart.

3. *Tight-fitting clothing obstructs the natural flow of the blood* by pressing upon the blood-vessels. The great veins which carry the blood from the head to the heart lie very near the surface in the neck, and, when clothing is worn tightly about it, the flow of blood is obstructed, and congestion of the veins of the brain may result.

4. At the junction of the chest with the abdomen are located the lower portions of the lungs, the stomach, the liver, and here the aorta branches off into several large blood-vessels. It is of the greatest importance that the action of these organs and the circulation of the blood be not hindered at this point by tight clothing.

5. Influences which tend to unduly obstruct the circulation of the blood and the free action of the organs must always result in *disease, and shortening of life.*

QUESTIONS

FOR

EXAMINATION AND REVIEW.

QUESTIONS.

PART III.

THE BLOOD.

Lesson I.

(a) — 1. What is the blood?
    2. Where is the blood found?
    3. What is the average quantity in the human body?

(b) — 1. How is the blood composed? What is the size of the corpuscles?
    2. Describe the weight of the corpuscles, and state how they congregate.
    3. Tell what you know of the composition of the plasma. *Note.* How do corpuscles of human blood compare in size with those of the blood of the lower animals? Of what use has a knowledge of this fact been in murder-trials?

Lesson II.

1. What materials does the blood contain?
2. In what materials is the plasma rich?
3. What are the corpuscles, and what do they contain?
4. What office does the blood perform? What does it convey to, and what remove from, the organs of the body?
5. State what you know of the motion of the blood during life. What are these movements called?
6. Name the organs of the circulation.

Rem. — By whom and when was the circulation of the blood discovered? What effect will the injection of blood from
the veins of one animal into those of another that has
lost much blood have? What is this operation called?
Has the operation been practised upon man? Are such
operations now performed?

Lesson III.

(a) — **1.** What is the heart? Where is the heart located?
(b) — **1.** Describe the form and size of the heart.
    2. In what is it enclosed? What is the name of this sack?
Describe it.
    3. Into what is the heart partitioned? Name the upper
chambers. The lower chambers.
    4. How are the chambers connected? Which sides of the
heart do not open into each other?
    5. Describe the walls of the ventricles. Why are the walls
of the ventricles so thick?
    6. Why do not the auricles need walls as strong as those of
the ventricles?
    7. Describe the opening (valve) between the right auricle
and the right ventricle, and give its name. Describe
the valve between the left auricle and the left ventricle,
and give its name. Describe the openings outward
from the ventricles, and give their names.

Lesson IV.

(c) — **1.** What motions has the heart? What occurs during its
contraction? What when it expands?
    2. What occurs when the right auricle contracts? What
when the right ventricle contracts? Through what
artery does the blood pass from the heart to the lungs?
    3. What happens to the blood in the lungs, and what be-
comes of it? What is the effect of the contraction of
the left auricle? Of the left ventricle? What then
conveys the blood to all parts of the body?
Rem. — How is the heart itself supplied with nourishment? What
is the "beating of the heart"?

Lesson V.

(a) — **1.** What are the arteries ? Where situated ?
 2. Where are the large arteries located ? What advantage
  does their depth give them ? Near what are many of
  them found ?
(b) — **1.** What does the arrangement of the arteries resemble ?
  Explain why.
 2. Of what are the arteries composed ? Describe the internal
  coat. The middle coat. The outer coat. By what are
  the coats nourished ?
(c) — **1.** Describe the action of the arteries. In what does this
  action aid ? What kind of blood do the arteries carry ?
Rem. — Describe the flow of blood from an artery that has been
  cut. From a vein.

Lesson VI.

(a) — **1.** Where are the capillaries located ?
(b) — **1.** Why is it impossible to tell where an artery ends, and
  where a vein begins ? How are the capillaries constructed ? What is their size ?
 2. How closely are they placed ?
(c) — **1.** What work do the capillaries perform ?
 2. What do they receive from the corpuscles ? What do
  they give up in return ? What two organs do they
  serve ?

Lesson VII.

(a) — **1.** Where do the veins begin ? Describe their general location.
(b) — **1.** What can you say of the thickness of the walls of the
  veins ? Of what are they composed ?
 2. With what are the inner coats provided ? How are these
  valves constructed ?
 3. What is the size of the veins at first ? What takes place
  soon after they leave the capillaries ? Describe their
  final union. To what does the final great vein extend,
  and how ?

(c) — **1.** What may the veins be called? What work do they
perform? *Note.* What does the great vein receive
just before it empties into the heart?

Rem. — Why is it difficult to ascertain how rapidly the blood
flows through the system? How quickly does the
entire blood pass through the heart?

Lesson VIII.

1. Where does the blood go from the right auricle?
2. Where does it go from the right ventricle?
3. Where from the lungs?
4. Where from the left auricle?
5. Where from the left ventricle?
6. Where from the arteries?
7. Where from the capillaries?
8. Where from the veins?

Lesson IX.

1. How frequently does the heart of an adult beat?
2. How does digestion influence the action of the heart?
Alcohol, etc.?
3. How does mental labor affect the action of the heart?
4. What is said of the effects of muscular exercise and of
violent efforts?
5. How is the circulation affected by fever?
6. What is said of causes which unduly excite, and produce
prolonged increase of, the heart's action?

Lesson X.

1. What influence has sleep on the action of the heart?
2. How may increased action of the heart be avoided in
special instances?
3. Tell what you know of the bad effects of tight-fitting
clothing. Of tight clothing about the neck.
4. What organs are situated at the junction of the chest and
abdomen? What sometimes interferes with the work
of these organs?
5. What is said of habits that tend to unduly hasten the
circulation?

PART IV.

THE

BREATHING APPARATUS.

" Fly, if you can, these violent extremes
Of air: the wholesome is nor moist nor dry.
But as the power of choosing is denied
To half mankind, a further task ensues, —
How best to mitigate these fell extremes,
How breathe unhurt."

THE BREATHING APPARATUS

˙ Lesson I.

(a) — **1.** The organs of respiration or breathing are, —
1. The *larynx.*
2. The *trachea.*
3. The *lungs.*

NOTE. —The passages of the nose and the mouth may be considered as the outer openings of the breathing apparatus.

EXPLANATION OF FIG. 24.

This figure gives a front view of the *larynx,* or vocal box.

The bone at the root of the tongue is seen, like half of a hoop, marked 8.

The front of the *thyroid cartilage* (Adam's apple) is marked 2.

3, the *cricoid cartilage.*

5, 6, other cartilages.

7, 7, are ligaments that suspend the bone of the tongue.

9, a ligament which connects the bone of the tongue with the Adam's apple, or thyroid cartilage.

The windpipe (*trachea*) is the tube at the bottom of the larynx.

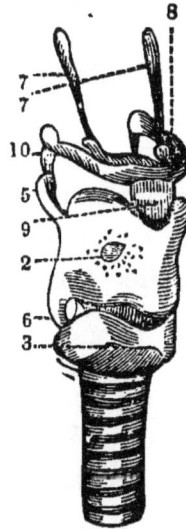

FIG. 24.

(b) **Position.** — **1.** The *larynx* is situated at the upper end of the windpipe, just behind the tongue.

2. In the front of the neck there is a bulge, which changes its place when we swallow. This bulge is frequently called Adam's apple. It is the front of the larynx.

(c) **Construction.** — **1.** The larynx is a small, muscular bulb. It is really an expansion of the upper end of the *trachea*, or windpipe.

2. It is a complex piece of mechanism, resembling a box composed of pieces of cartilage which may be moved on each other, and which enclose bands of membrane, — the *vocal cords*, — by whose vibrations voice is produced. The cartilages of which the skeleton of the larynx is composed are five in number.

3. The larynx opens into the back chamber of the mouth by a narrow chink called the *glottis*. This opening is provided with a small, spoon-shaped lid called the *epiglottis* (*epi*, upon, *glotta*, the tongue).

(d) **Work.** — **1.** The glottis is usually open; but the epiglottis is placed there to cover the opening when the food passes over it on its way to the œsophagus. When the act of swallowing is not taking place, this valve stands open for the free admission of air into the trachea and lungs.

2. The larynx is the organ of voice. On each side of the glottis elastic membranes project from the sides of the larynx across the opening. These membranes are called "vocal cords," and, when not in use, they spread apart and leave an angular opening, through which the air passes into and out of the lungs, without producing voice.

3. When we wish to use the larynx in producing voice, the muscles attached to this wonderful instrument tighten the vocal cords, draw them parallel with each other, and very close together. The passage of a current of air between the parallel edges of the vocal cords, sufficiently strong to set them vibrating, produces voice. Sounds are thus produced in the same way as by the rapid vibrations of the "tongues" of the accordeon, or the strings of the violin.

4. Certain muscles which are attached to the glottis draw its sides more closely together, or allow them to separate, and, by thus narrowing or widening this chink, the sounds made are varied and modified.

Remarks.—The larynx is but slightly developed in early infancy, and does not at this period of life differ in size in the two sexes, nor does the character of the voice of the two differ at this time. This organ remains nearly stationary in size from the third to the twelfth year; but about the fourteenth year it almost doubles in size in the boy, and the voice takes a masculine tone. This change is rapid, and is nearly completed in the course of a year, though the larynx is not fully developed till about the twenty-fifth year. In girls, it increases about a third in size; and the larynx of a woman is smaller than that of man. These differences in size account for the compass and power which distinguish the voice of man from that of woman.

The tones of the voice, usually, will be high or low according as the vocal cords are tightened or relaxed. When, however, the larynx is diseased by what is known as a *cold*, the natural tones are changed, and frequently destroyed, and the afflicted person can only speak in a whisper.

If we laugh, talk, or attempt to breathe when we swallow, the epiglottis opens slightly, and allows particles of food to "go the wrong way," that is, into the larynx; and this organ endeavors to expel the trespassing particles by a violent expulsion of air from the lungs. Such an expulsion of air is called a *cough*.

Lesson II.

THE TRACHEA OR WINDPIPE.

(a) *Position.* — 1. The *trachea* (*trachea*, rough) is a tube which extends from the larynx downward to the lungs, in front of the œsophagus, and parallel with it.

2. The rings of cartilage which form the skeleton of the trachea are easily felt in front of the throat.

(b) *Construction.* — 1. The trachea is composed of stiff, parallel rings of gristle ensheathed in a tough membrane. The rings of gristle strengthen and keep it open for the passage of air.

2. At its lower end the trachea divides into two branches called *bronchi*, one of which leads to the right lung, and the other to the left lung.

3. After entering the lungs, these branches again divide into smaller ones, and these into still smaller

ones, like the branches and twigs of a tree ; and the
tiny branches of these tubes at last end in clusters
of air-cells, which are only about $\frac{1}{100}$ of an inch in
diameter.

4. These air-tubes, large and small, are lined with
an extremely delicate, silky lining called *mucous membrane.*

(c) *Work.* — 1. The trachea and its branches convey the air to the lungs, carrying it into the minute
air-cells, in which it gives up its oxygen and becomes
charged with gases set free from the blood, when it
is conveyed out of the body by the same air-tubes.

Remarks. — The lining of the air-tubes is so extremely
delicate and sensitive, that, while it will bear the presence of
pure air, it will not permit the touch of any other substance,
not even that of a drop of pure water. Indeed, it may be
safely asserted that carbonic acid gas when undiluted cannot
be breathed. The epiglottis, like the faithful pylorus of the
stomach, stands sentinel, closes the door, and forbids the passage into the lungs, for the time at least, of a gas so deadly.

Most of our colds and coughs are results of irritation or
inflammation of the mucous membrane of ·the air-tubes and
lungs.

Lesson III.

THE LUNGS.

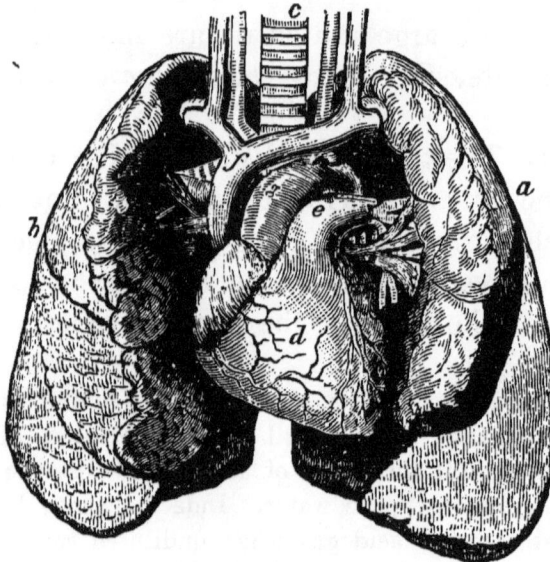

EXPLANATION OF
FIG. 25.

a, the left lung.
b, the right lung.
c, the windpipe.
d, the heart.
e, the great artery carrying blood to the lungs.
f, the great vein.
g, the great artery carrying blood to the body.

FIG. 25.

(a) *Position.* — 1. The *lungs* are situated within the chest, one on the right side and the other on the left, with the heart between them.

(b) *Construction.* — 1. The lungs are very soft, spongy, and elastic, contain but little flesh, and are mainly composed of small tubes and air-cells.

2. In shape, the lungs are irregular cones, resting on their bases, and supported from beneath by the diaphragm.

3. The substance of the lungs is of a grayish rose-color. They are surrounded by a double sack, the

pleura, one layer of which is attached to the lungs, and the other to the walls of the chest. It gives out a fluid which oils its surface so that one layer may glide upon the other with such perfect ease that the lungs are well protected from injury in coming in contact with the walls of the chest.

4. The lungs are not muscular, and therefore have no power to act for themselves in respiration.

(c) *Work.*—**1.** The office of the lungs is to supply the blood with *oxygen*, and to remove the *carbonic acid gas* and watery vapor which should be cast out from it.

2. On entering the air-cells of the lungs, the air is separated from the blood, which has been sent to the lungs from the heart for purification, by the thin walls of the cells only.

3. The oxygen of the air passes through the pores of the walls of the air-cells, combines with the impure blood, which is of a dark color, changes it to a brilliant red. While the air gives up its oxygen to the blood, it receives in return carbonic acid gas, watery vapor, and other impure waste-matter, with which the blood has become charged in its journey through the system ; and these are cast out from the lungs at every expiration (*breathing out*).

4. Charged anew with the life-giving oxygen, and relieved of poisonous gases and worn-out particles, the pure, red stream flows back to the heart to be sent again on its mission to all parts of the system, carrying nourishment to them.

Remarks. — Common air is mainly composed of two fluids, *oxygen* and *nitrogen*, there being about twenty-one parts of oxygen to seventy-nine parts of nitrogen. Oxygen enters into the composition of all animal and vegetable matter, and is constantly necessary to life in all its various forms, — to the ponderous elephant and the tiniest insect, to the immense tree of the forest and the smallest blade of grass.

When the air enters the lungs, the blood absorbs some of its oxygen. Air, therefore, which has been breathed one or more times, has lost much of its oxygen, and has become heavily charged with foul gases from the blood. It is no longer fit to be breathed.

Lesson IV.

HOW WE BREATHE.

(a) *Respiration Defined.* — **1.** *Respiration*, or breathing, consists of two movements, — *inspiration* and *expiration*.

2. *Inspiration* is the drawing of air into the lungs.

3. *Expiration* is the expelling, or forcing out, of the air from the lungs.

(b) *Inspiration Described.* — **1.** In taking a full breath, we throw the chest forward, the shoulders back, and straighten the backbone. This is done in order to give free play to the muscles that move the breathing apparatus.

2. The *diaphragm* (a broad muscular partition between the chest and the abdomen) lowers itself so as to press the walls of the lower part of the chest outward, and increase the size of the cavity.

3. On being relieved from all pressure, the elastic lungs expand, and air rushes in through the nostrils, trachea, and bronchial tubes, and fills the vacant air-cells of the lungs.

(c) *Expiration Described.* — 1. The diaphragm is pushed upward against the lungs by the contraction of the muscles of the abdomen, the walls of the chest contract, and the ribs are pulled downward by their muscles.

2. The size of the chest is greatly diminished by these movements, and the air is pressed out of the lungs through the air-tubes, bronchi, larynx, and nostrils.

(d) *Frequency of Respiration.* — 1. In an adult in a condition of repose, respiration takes place about eighteen times a minute. In the infant it is more frequent. Respiration becomes very active under the influence of bodily exercise or under excitement of the mind. But, on the other hand, when the attention becomes fixed in laborious mental effort, the breath is held, so that it soon becomes necessary to take long, deep inspirations to "make up" (compensate) for the small supply furnished the lungs by the preceding inspirations. This insufficient respiration should be guarded against by students when long employed in mental effort, as their constitutions suffer greatly from a lack of supply of air in the lungs.

2. As the air is less dense in elevated regions than in the lower regions, on the seacoast for example, it is necessary to breathe oftener upon high mountains in order to supply the lungs with the required amount of oxygen. But this increase of respiration is noticeable only when the height is considerable, and the distance is passed over rapidly.

———

Lesson V.

THE AIR WE BREATHE.

(a) *Pure Air Needed.* — **1.** Whether the blood shall carry the nourishing and life-giving oxygen to every part of the body, or whether it shall return to the heart from the lungs without being relieved of carbonic acid, depends entirely on the purity of the air we breathe.

2. *Pure air makes pure blood;* and pure, rich blood gives nourishment to all the organs; but impure air — air that contains but little oxygen, and is laden with carbonic gas and other impurities — poisons the blood with carbonic gas, and starves it for want of oxygen.

(b) *The Air of Close Rooms Poisonous.* — **1.** In a short time the air of a close room becomes filled with carbonic acid and other matter thrown out of the body through the lungs and skin, while the oxygen has all been consumed. In this condition it is poisonous, and unfit to be breathed. If a room were perfectly air-tight, or even nearly so, the air would become so poisonous as to cause death.

2. Great care should be taken to admit a full supply of fresh air into all apartments of our houses, *particularly into our sleeping-rooms.* A simple way to do this, and at the same time prevent "taking cold" from a draught, is to insert a board six or eight inches wide, and long enough to reach entirely across the window, and fill the space under *the raised lower sash.* Currents of air will then enter only between the upper and the lower sashes, and will be projected upward, losing their force ere they reach the person of any one in the room.

(c) *Poisonous Air from Drain-Pipes.* — 1. If the drain-pipes leading from houses to cesspools and sewers are not constructed properly, poisonous carbonic acid and other foul gas will be carried back through them, poisoning the air of apartments, and causing disease and death.

(d) *Malaria.* — 1. *Malaria (bad air)*, as its name indicates, is a disease caused by breathing air that is filled with poisonous particles that arise from drain-pipes, decaying vegetable matter, marshy land, etc. It is supposed that little atoms, called *spores*, float in the air from these sources, and that they are absorbed into the blood through the lungs, and pores of the skin. These *spores* irritate and poison the blood, and create the disorder from which so many people in certain sections of our own country, and in warm countries, suffer.

(e) *Foul Air Causes General Disorder.*—1. Whatever deprives the lungs of the supply of oxygen required for the purification of the blood, or prevents them from casting out carbonic acid gas, sows the seeds of disease. It follows, then, that if the lungs are not only hindered in their work by improper clothing, but compelled to breathe poisonous air, disease will attack the weakest organs of the body first, and extend from these to others, finally ending in death.

QUESTIONS

FOR

EXAMINATION AND REVIEW.

QUESTIONS.

THE BREATHING APPARATUS.

Lesson I.

(a) — 1. Name the organs of breathing. What may be considered as the outer openings of the breathing apparatus?

(b) — 1. Give the position of the larynx. Where may the front of the larynx be felt? What is it commonly called?

(c) — 1. What is the larynx?

    2. What kind of a piece of mechanism is it? What does it resemble, and of what is it composed? What are the vocal cords?

    3. Into what, and how, does the larynx open? What is the glottis? The epiglottis?

(d) — 1. What is the usual condition of the glottis? What is the work of the epiglottis?

    2. Of what is the larynx the organ? Describe the position of the vocal cords. What is their position when not in use?

    3. How do the muscles of the larynx act when we wish to produce voice? How is voice then produced?

    4. How is voice modified and varied by the glottis?

Rem. — What is said of the larynx in infancy in the two sexes? How long does it remain nearly stationary in size? When does it increase to nearly double its former size? What change of voice does this increase cause? When

117

is the larynx fully developed? What is the extent of increase of the larynx of girls? How does the larynx of woman compare with that of man? For what do these differences in size account? What usually causes the voice to be high or low? What is the effect of a "cold" on the tones of voice? What occurs if we attempt to laugh, talk, or breathe while swallowing food or drink? What endeavor does the breathing apparatus make in such instances? What is such an effort called?

Lesson II.

(a) — 1. Where is the trachea, or windpipe, situated?
     2. Where may its skeleton be felt?
(b) — 1. Of what is the trachea composed? Of what use are the rings of cartilage?
     2. Describe the lower end of the trachea. What are these branches called, and to what do they lead?
     3. Describe the branches of the trachea after entering the lungs. In what do the tiny branches end? How large are the air-cells of the lungs?
     4. With what are the air-tubes lined?
(c) — 1. What is the office or work of the trachea and its branches? What occurs to the air after it enters the air-cells of the lungs?
Rem. — What is said of the nature of the lining of the air-tubes? Can we breathe undiluted carbonic acid gas? What is remarked of the action of the epiglottis when this gas attempts to enter the air-passages? Of what are most of our colds and coughs the results?

Lesson III.

(a) — 1. Where are the lungs located? How is the heart situated in regard to the lungs?
(b) — 1. What is the nature of the material of the lungs? Of what are they mainly composed?
     2. Of what shape are the lungs? On what do they rest?

3. What is the color of the lungs? By what are they enclosed? To what are the layers of the pleura attached? Of what use is the pleura? What does it give out?

4. Are the lungs muscular? Do they possess the power to act for themselves in breathing?

(c) — 1. What is the office or work of the lungs?

2. What partition separates the air from the blood in the lungs?

3. How does the oxygen of the air reach the blood? What change does it make in the impure blood? What does the air in the lungs receive from the blood? What becomes of these impurities?

4. With what does the blood in the lungs become charged? Whither is the blood sent from the lungs?

Rem. — Of what is common air mainly composed? In what proportions of each? What is said of the nature of oxygen, and into the composition of what does it enter? With what does the air in the lungs part? What is the condition of air that has been breathed one or more times? Should it be breathed in this condition?

Lesson IV.

(a) — 1. Of how many and what acts does respiration consist?

2. What is inspiration?

3. What is expiration?

(b) — 1. Describe the action of the spine and shoulders when we take in a full breath. Why do they perform these actions?

2. Describe the action of the diaphragm during inspiration. What effect has this action?

3. Describe the action of the lungs in inspiration. What does the air then do?

(c) — 1. Describe the action of the muscular diaphragm in expiration. Of the walls of the chest, and of the ribs.

2. What is said of the size of the chest in expiration? What effect have these actions upon the air in the lungs?

(d) — 1. How often does an adult in repose breathe during a minute? How often does an infant breathe? What effect do bodily exercise and excitement of the mind

have on respiration? What effect does fixed and laborious mental effort have on respiration? Why do
we take long, deep breaths after the breath has been
held in that way? What habit of respiration should
students guard against? Why? Why is it necessary
to breathe oftener in elevated regions than in lower
regions? When is this increase of respiration noticeable?

Lesson V.

(a) — 1. Upon what does the purity or impurity of the blood
greatly depend?

    2. What does pure air make? What is impure air, and what
effects has it on the blood?

(b) — 1. With what does the air of close rooms become filled?
What is said of such air?

    2. What care should be taken of the air of all apartments?
What is an easy way to admit fresh air without
draughts?

(c) — 1. What is said of poorly constructed drain-pipes?

(d) — 1. What does the word "malaria" mean? What causes the
disease called by this name? State what is said of
spores. What effect have they when absorbed into the
blood?

(e) — What is the effect of depriving the lungs of a proper
supply of oxygen? If the lungs are deprived of the
necessary oxygen, what organs will disease attack
first? Which next?

PART V.

THE MUSCLES.

———————

"Each *fibre* ranged with such amazing skill
That every *muscle* may attend thy will,
How every *tendon* acts upon its *bone*,
And how the *nerves* receive their nicer tone."

THE MUSCLES.

Lesson I.

(a) *What Muscles are.*—1. *The muscles are the instruments of motion.* While the body owes its general form to the bones, its power of motion and its beautiful proportions are given by the muscles.

2. The *muscles* and *tendons* are to the human body what the ropes and sails are to the masts and spars of a ship. As a ship without sails and ropes would be a very unmanageable thing, so the body without muscles and tendons would have no power to move, or direct its position.

3. In the bones of the body we find the *columns*, *levers*, and *pulleys* of a complex machine; and in the muscles and tendons we have the *cords*, *belts*, or *springs*, which move the bony levers and pulleys.

4. The muscles of an animal body are the *lean* meat. Lean beef, the deep-red flesh of the cow or ox, is the muscular part of the animal's body. There are more than five hundred muscles in the human body.

123

(b) *Position.* — 1. The muscles are situated in all parts of the body. The great mass of flesh covering the skeleton is mainly composed of them, while the organs situated in the cavities of the body are either muscles, or have muscles connected with them. Among the muscles situated within the framework are the heart, the diaphragm, the muscular coat of the stomach, and the tongue.

(c) *Construction.* — 1. The muscles are composed of fine *fibres* or strings held together by a connecting network of tissue, and bound up in smooth, silky casings of thin membrane. The microscope enables us to see that each of these fibres or strings is formed of still finer ones.[1]

2. The muscles are laid one over the other, separated by the smooth casings of membrane, and by layers of fat that enable them to move without interfering with each other. These layers of fat give a plumpness of form which the body would not otherwise have.

3. In shape and in length, the muscles vary greatly. Some are round; others flat, square, or triangular. Some of the muscles of the larynx are only about one-eighth of an inch in length, while the *sartorius*, or "tailor's muscle," by which the legs are crossed, is nearly three feet in length.

4. Muscles are large and thick in the middle, but

---

[1] The muscular fibres are readily separated in a piece of boiled meat.

small at the ends. The middle part is called the body, or *swell*, and it possesses the power of contraction. The extremity of the muscle attached to the bone which is moved is called the *insertion*, or free end of the muscle: the extremity towards which it draws in contraction is called the *origin*, or fixed end of the muscle. Generally the *origin* of a muscle is nearest the trunk.

EXPLANATION OF FIG. 26.

In this figure the biceps muscle is shown at C, and the two tendons which attach it to the shoulder are seen at G, the point of origin.

The attachment of the muscle to the radius is shown at A, the point of insertion.

FIG. 26.

THE BONES OF THE UPPER EXTREMITY AND THE BICEPS AND TRICEPS MUSCLES.

The triceps muscle is represented at F, and the tendon by which it is attached to the radius is shown at B. These two muscles are *antagonistic* muscles.

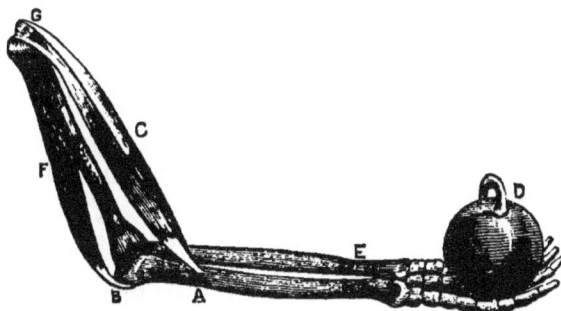

5. At the ends, the threads or fibres of the muscle change into strong, tough *tendons*, of a bluish-white color, which are firmly fastened to the bones. The tendons have no power of contraction, and are merely the ropes, as it were, by which the body of the muscle is fastened to the bone, or other part, which is moved by the contraction.

6. The muscles vary in color. The weaker ones are usually of a pale rose-color, while the larger and stronger ones are deep red. The color of the muscles deepens when they are exercised.

7. At least one artery enters each muscle, and supplies it with blood for its nutrition. A nerve also penetrates each muscle, and connects it with that great central office of the nerves, the brain, so that it may be subject to the will.

Lesson II.

WORK OF THE MUSCLES.

1. All movements of the different parts of the body are caused by the contraction of muscles.

2. The cells which compose the muscles are peculiarly elastic, and have the power to widen out, making each fibre of the muscle shorter and thicker. This power of these cells is the source of the contraction of the muscles which produces all bodily movements. The contracting muscle shortens and thickens, and pulls the movable part to which it is attached with it. A good illustration of this action is found in the work of the muscles that bend the arm. (*See Fig. 26.*) The *biceps* muscle contracts, and pulls the bones of the lower arm upward, toward the shoulder: the *triceps* contracts, and pulls the bones of the lower arm back again, thus straightening the arm. If both of these muscles contract

at the same instant, there can be no movement of
the elbow-joint, and thus we see the antagonistic
nature of these two muscles.

3. More than two hundred muscles are arranged
in pairs, one to draw a part in one direction, and the

EXPLANATION OF FIG. 27.[1]

a, the *occipito-frontalis*,
the muscle that raises the
eyebrows and skin of the
forehead.

c, the *orbicularis palpe-
barum*, to shut the eye.

n, the *levator palpebræ
superioris*, to open the eye
by raising the upper lid.

p, the *corrugator super-
cilii*, to wrinkle the eyebrows.

o, the *compressor nasi*,
to compress the wings of
the nose.

d, the *levator labii superioris*, to pull the upper lip
directly upward.

e, the *zygomaticus minor*,
to raise the corner of the
mouth outward.

f, the *zygomaticus major*,
to swell the cheek, and raise
the corner of the mouth.

FIG. 27.

i, the *orbicularis oris*, to shut the mouth by contracting the lips.

k, the *depressor anguli oris*, to draw down the corner of the mouth.

l, the *levator anguli oris*, to raise the corner of the mouth.

m, the *depressor labii inferioris*, to draw the under lip downward and outward.

h, the *masseter*, to raise the lower jaw.

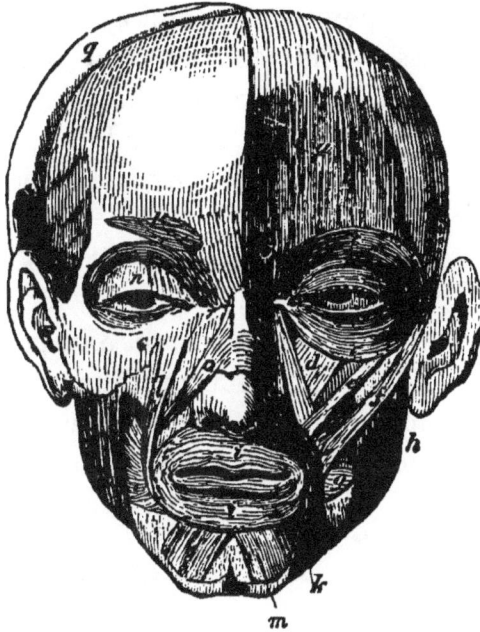

[1] Each muscle has a name, which has been given to it because
of its size, shape, or the work which it performs. The names are
here given to satisfy curiosity, rather than for memorization.

other to restore it to its former position, or to hold it motionless at any required point in the range of its motion. These pairs are called *antagonists.*

4. All muscles do not move bones, and bend joints; but some have quite different work to perform. The heart, which is a compound muscle, exerts its contractile powers in forcing the blood through the arteries. The stomach and other muscles of the digestive organs exert their force in mixing, churning, and moving the food in preparing it for the nourishment of the body; and the muscles of the eye move that organ.

5. In the human face, all the various expressions that indicate the emotions of the mind — joy, sorrow, hatred, affection, pleasure, and pain — are caused by the contraction and swelling of the muscles which produce the lights and shadows of the countenance. Reference to *Fig. 27* will assist in forming an idea of some of the principal muscles of the face, and of their work.

Lesson III.

CLASSES OF MUSCLES.

1. The muscles are divided into two great classes; viz., *voluntary* and *involuntary.*

(a) *Voluntary Muscles.* — 1. The voluntary muscles are those that are under the control of the will. They move, or cease to move, when the mind wills it.

The muscles of the fingers, limbs, trunk, and many others, belong to this class.

(b) *Involuntary Muscles.* — **1.** The involuntary muscles are those that act independently, and are not under the control of the will. The muscles of the stomach, heart, and those that move in sneezing, coughing, and shivering as from a chill, are among the muscles of this kind.

2. The movements of certain muscles appear to be involuntary, but are not really so. The sudden winking of the eye when any object threatens it may be considered voluntary; for, if the attention is attracted, the will can control the movement.

(c) *Flexors and Extensors.* — **1.** The muscles that bend a joint, or move any part, are called *flexors:* those that restore the parts to their former position are called *extensors.* Every joint in the body is provided with at least one pair of these muscles. At the elbow-joint (*Fig. 26*) the biceps muscle contracts, and bends the joint, and a contraction of the triceps muscle straightens it again.

2. Other muscles of these kinds produce a twisting or rolling motion of a limb, as in the fore-arm when the palm of the hand is turned upward or downward.

3. These flexors and extensors are attached very obliquely to the bones, and do not exercise as much power as they would if placed more nearly at right angles. Such an arrangement would interfere with convenience and beauty of form, and freedom of

movement.   The limbs would be unwieldy and
nearly useless.

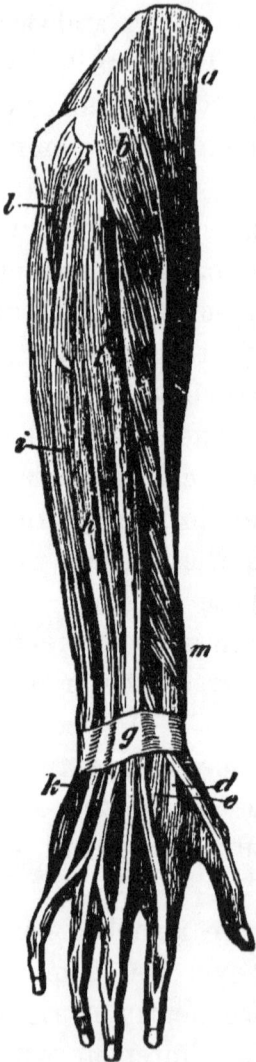

EXPLANATION OF FIG. 28.

f, the muscle that straightens the fingers.

h, the muscle that straightens the little finger.

i, the muscle that assists
in straightening the wrist.

l, the muscle that assists
in extending the fore-arm.

d, the muscle to extend
the second bone of the
thumb outward.

e, the muscle to extend
the fore-finger.

k, the muscle to draw the
little finger outward.

m, the muscle to roll or
turn the fore-arm, and turn
the hand.

g, the ligament which
binds down the muscles at
the wrist.

———

EXPLANATION OF FIG. 29.

a, the muscle to turn the
hand inward.

b, the muscle to bend the
wrist.

c, d, the muscles to bend
the hand.

e, the muscle to assist in
bending the hand.

g, the muscle to bend the
thumb.

FIG. 28.

FIG. 29.

4. Strong bands of ligament bind down the muscles, keep them in place, and add to their strength. The muscles at the wrist and ankle are thus firmly held in place, and prevented from flying from the bones when strongly contracted. *Figs. 28* and *29* present front and back views of the fore-arm, in which the long flexor and extensor muscles are represented.

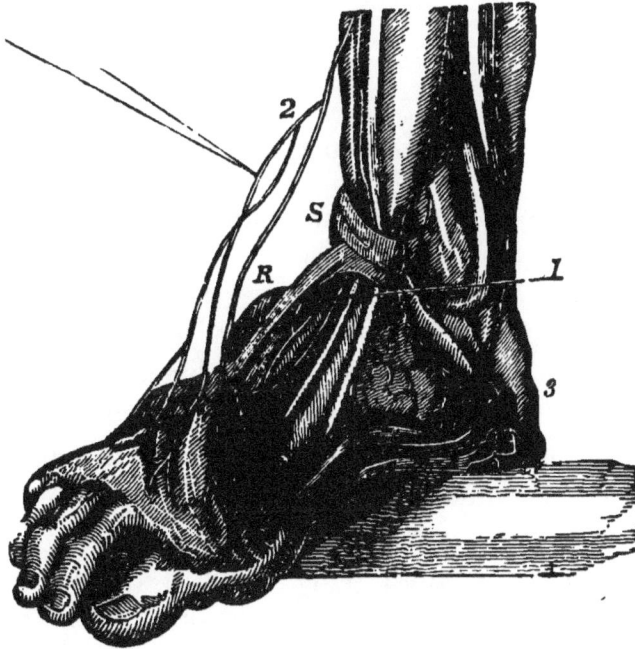

FIG. 30.

EXPLANATION OF FIG. 30.

S, the ligament that binds down the muscle to the bone of the ankle.

R, a muscle which assists in extending the toes.

1, tendons of the muscles that move the toes.

2, nerves of the toes, pulled upward.

3, at the back of the ankle, the tendon of Achilles, which extends from the muscle of the calf to the back of the heel. This is the strongest tendon of the body : it raises the heel.

5. In the fore-arm, between the elbow and the ends of the fingers, there are about fifty muscles. In the fingers there are numbers of short, delicate ones, capable of very quick movement: these have been called "fiddler's muscles." In the figures of the arm may be seen the tendons that proceed from the muscles to the bones of the fingers. The arrangement of the muscles of the foot is similar to that of the hand.

———

Lesson IV.

THE SOURCE OF MOTION. — EXERCISE.

(a) *Why Muscles Contract.* — **1.** The muscles receive their power from the brain and nerves.

2. Each muscle is penetrated by a nerve which connects it with the brain, or spinal marrow. This nerve branches out, and sends tiny threads into the fibres of the muscle, and in this respect each fibre is separate from every other.

3. When the mind wills to move a muscle, the brain sends out a mysterious agent through the nerves to the cells of each fibre of that muscle, and it contracts. This is all that is known of that strange stimulus by which bodily movement is directed.

4. If a nerve be cut anywhere between the spinal cord and the muscle to which it belongs, the muscle instantly loses its power of motion. This clearly proves that the nerves convey to the muscles a power which they have not within themselves.

(b) *Exercise of the Muscles.* — 1. It is a general law of the body, that exercise is necessary to the health of all its parts. Tie up a blood-vessel, and it becomes a. withered, useless thing. The bones become weak, and dwindle away, when deprived of exercise; and so it is with the muscles. Lack of proper exercise causes softness, weakness, and inability to perform the work for which they are designed. This is not only true of the muscles that bend the joints, and move the limbs, but also applies to those employed in breathing, and to the vocal muscles.

2. Care must be taken that exercise be not too severe, nor continued so long as to produce exhaustion: muscle is weakened, rather than strengthened, by undue exertion.

(c) *Exercise Aids the Circulation.* — 1. When a muscle contracts, some of the veins are compressed, so that the blood cannot flow freely onward, and the valves of the veins forbid a backward flow. The arteries continue to force the blood along, and the veins become swollen. As soon as the contraction of the muscle ceases, the blood rushes onward with greatly increased speed.

2. Now, when a number of muscles are employed in strong, quick action, many veins are affected in this way, and the whole circulation is quickened. The heart must work faster to send the blood to the

lungs, and the lungs must work quicker to supply the oxygen required by the greater quantity of blood sent to them. The purified blood is carried back to the heart with greater speed, and the heart again forces it rapidly out through the arteries and capillaries to perform its mission.

(d) *Exercise Aids Appetite and Digestion.* — 1. When the blood reaches the capillaries, the quickened flow causes them to do their work faster, and the worn-out matter is removed more quickly. The organs call for new material, and the stomach demands more food to supply new blood to the system. Thus it will be seen that muscular exercise gives vigor to every part of the body.

(e) *Hints about Exercise.* — 1. *Exercise should be taken in pure air;* it calls for a full supply of oxygen to satisfy the increased demand.

2. *Exercise should not be taken just before nor soon after severe mental labor, nor immediately after a hearty meal.* In this latter instance the stomach requires the blood which would thus be called away from it, and delay its work.

3. *Tight clothing* interferes with the action of the diaphragm and other muscles used in breathing; and *tight shoes* interfere with the free movement of the muscles of the feet and legs, causing ungraceful and constrained action of these members

QUESTIONS

EXAMINATION AND REVIEW.

QUESTIONS.

PART V.

THE MUSCLES.

Lesson I.

(a) — **1.** What are the muscles? What gives the body its power and beauty of form?

2. To what may the muscles and tendons be compared? What comparison of the movements of a ship is made with those of the human body?

3. To what parts of a machine are the bones similar in their uses? To what are the muscles and tendons equivalent?

4. What part of the flesh is muscle? How many muscles are there in the human body?

(b) — **1.** Where are the muscles situated? Of what is the great mass of flesh composed? Mention certain muscles that lie within the cavities.

(c) — **1.** Of what are the muscles composed? How held together, and by what incased?

2. How are the muscles laid? What do the layers of fat give to the body?

3. What is said of the shape and length of the muscles? Mention the shapes, and state the sizes of certain muscles.

137

4. In what part is a muscle thick? Thin? What is the middle part called? Give the names applied to the ends of a muscle. Where generally is the origin of a muscle?

5. Describe the structure of the ends of a muscle. What is the nature of the tendons? What is the use of the tendons?

6. What is the color of the muscles? What effect has exercise on the color?

7. State how the muscles are supplied with blood-vessels and nerves.

Lesson II.

1. How are all movements of the body produced?

2. What is the nature and what the power of the cells of the muscles? Of what is this property of the cells a source? Describe the action of a muscle in producing movement of a part of the body. In what is a good illustration of this action found? Describe the work of the muscles that bend the arm. If both of these muscles contract at the same time, what is the result?

3. What is said of pairs of muscles? Describe their work. What are such muscles named?

4. Do all muscles move bones and bend joints? State what you know of muscles that have a different office.

5. How are emotions of the mind expressed in the countenance? *Note.* What is said of the names applied to muscles?

Lesson III.

1. Into what general classes are all muscles divided?

(a) — 1. What are voluntary muscles? Mention some.

(b) — 1. What are involuntary muscles? Mention some.

2. What is said of some muscles that appear to be involuntary?

(c) — 1. What are flexors? Extensors? With what is every joint provided?

2. What kind of motion do some of these muscles produce?

3. How are the flexors and extensors attached to the bones? What effect on the power has this position of the muscles? What would be the effect if they were placed more nearly at a right angle?
4. What bind down the muscles? What is said of the muscles of the wrist and ankle?
5. About how many muscles are in the fore-arm? What is the nature of the muscles of the fingers, and of what are they capable? What is said of the arrangement of the muscles of the foot?

Lesson IV.

(a) — 1. From what first source do the muscles receive their power?
2. By what is each muscle penetrated? Describe the branching of the nerve.
3. How does the mind move a muscle?
4. What is the effect of severing a nerve? What does this prove?

(b) — 1. What is a general law in regard to exercise? What is said of tying up a blood-vessel? What effect has lack of exercise on the bones and muscles? Is this true of all muscles?

(c) — 1. What effect on the veins has the contraction of a muscle? What happens when the contraction ceases?
2. When many muscles contract, what is the effect upon the circulation? What of the action of the heart? What of the speed of the blood?

(d) — 1. How are the capillaries affected by increase of circulation? The organs of the body? The stomach? What does all this show?

(e) — 1. Where should exercise be taken? Why?
2. What is said of exercise after mental labor? Why should we not engage in severe exercise after a hearty meal?
3. How does tight clothing interfere? Tight shoes?

PART VI.

THE BRAIN AND NERVES.

" These outguards of the mind are sent abroad,
And still patrolling beat the neighboring road,
Or to the parts remote obedient fly,
Keep posts advanced, and on the frontier lie."

THE BRAIN AND NERVES.

Lesson I.

THE BRAIN.

(a) *Position.* — **1.** The *brain*, the principal organ of intelligence, is situated in the head, and is surrounded and protected by the bones of the skull.

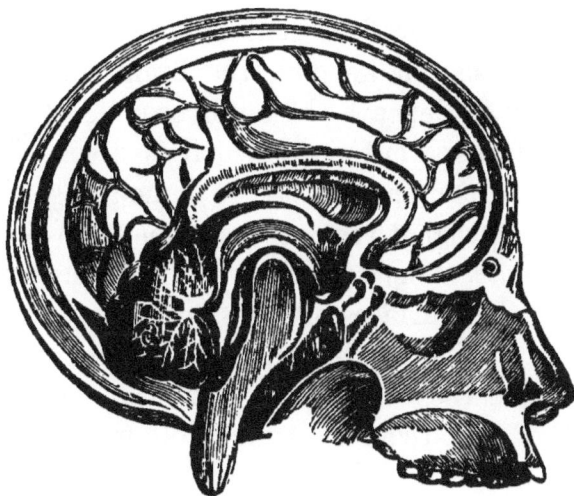

FIG. 31.

EXPLANATION OF FIG. 31.

This figure represents the skull as divided in the middle, from front to back, down to the neck.

2. The brain is divided into two parts. That which occupies the cavity of the skull above the level of the ears is called the *cerebrum*, or *great brain :* the part which fills the cavity below the level of the ears, at the back of the head, is called the *cerebellum*, or *little brain.* A membrane, tightly stretched, separates the two parts, and relieves the lower brain from the weight and pressure of the upper one.

(b) *Construction.* — **1.** When the bones are removed, a thick, shining membrane is seen. This is the *dura mater*, or firm coat of the brain, and its office is to assist in keeping the brain together, and to protect it. Beneath the outer coat lies the *arachnoides*, or transparent coat, which is a very delicate, transparent membrane. It so much resembles a spider's web, that it receives its name from that fact, — *arachnoides*, "the spider's web." This membrane lies over the surface of the brain, but does not closely follow its depressions. The third and inner coat is called the *pia mater*, or soft coat. It is a thin network of blood-vessels, which follows the fissures, and winds into the substance, of the brain.

2. The substance of the brain consists of two kinds of matter; viz., *gray* and *white.* The gray matter forms the outside of the brain, and the white the inner portions.[1] So extremely soft is the substance of the brain, that it would fall apart from its

---

[1] The gray is supposed to be the portion that originates a fluid which imparts power of motion ; and the white is supposed to conduct the fluid to all parts of the body.

own weight if it were not surrounded by its membranes.

3. The outer surface of the brain is not smooth and regular, but consists of worm-like ridges interspersed with hollows : in other words, it is furrowed.

(c) *Work of the Brain.* — **1.** The brain is the seat of thought, of intelligence, of sensation, and of motion. The knowledge which has been obtained concerning the special uses and work of the different parts of the brain is very limited, and mainly founded on supposition.[1]

2. It is believed that the *cerebrum* is the chief organ of the mind, and that it presides over the intellectual processes. It is there that we think, reason, and will.

3. Various kinds of work have been attributed to the *cerebellum ;* but one kind only has been generally admitted. Experiments seem to prove, that, if the *cerebellum* be injured or removed, a confusion of movement of the muscles is caused, like that produced by alcoholic intoxication. It is believed, therefore, that this organ is the regulator of muscular motion.

(d) *Peculiarities of the Brain.* — **1.** The brain suffers no pain from wounds. A portion of it may be cut off without creating pain. Portions of the brain sometimes escape through fractures of the skull, and

---

[1] Most of the theories in regard to the functions of the brain are disputed and uncertain, and physiology is very reserved in regard to them.

still the injured person recovers without suffering injury to his powers of mind.

2. If the upper part of the *cerebrum* of an animal be removed, he becomes blind and apparently stupefied, but may be roused, and then can walk steadily and naturally.

3. The *medulla oblongata* (that portion of the brain next to the spinal cord) is probably the most delicate and sensitive portion of the body. The slightest injury, the prick of a needle, to this organ causes instant death.

Lesson II.

THE NERVES.

(a) **Location of the Nerves.** — **1.** *Nerves* spring from the brain and spinal cord, and extend to every part of the body.

2. Certain nerves start from the base of the brain, within the skull, and extend to the eye, ear, tongue, nose, throat, stomach, heart, etc. These are named *cranial* nerves, because they begin in the *cranium* or skull.

3. The *spinal cord*, which is an extension of the substance of the brain, extends downward through the tube or canal of the backbone. Between the points of the bones of the spine, the spinal cord sends out branches, which are named *spinal nerves.* These extend to the arms, the chest, the abdomen, the legs, etc., and have various names.

4. The nerves branch out from the spinal cord precisely like the limbs and smaller branches of a tree. *Fig. 32* gives a view of the brain, spinal cord, and starting-points of the nerves.

(b) *Construction.*—1. The nerves are branches and twigs of the brain. They consist of the same substance as the brain, and, like it, are surrounded and protected by sheaths of membrane.

2. The nerves branch off in pairs from the brain and spinal marrow, through little openings in the bones. Twelve of these pairs spring directly from the brain, and thirty of them from the spinal cord, sending their branches and twigs to every muscle, blood-vessel, or other organ of the body.

3. Nerves are of all sizes, from one-fourth of an inch in diameter to hair-like threads, so small as to be invisible to the unassisted eye. In length they differ as much as in thickness. The brain, spinal cord, and nerves constitute the *nervous system.*

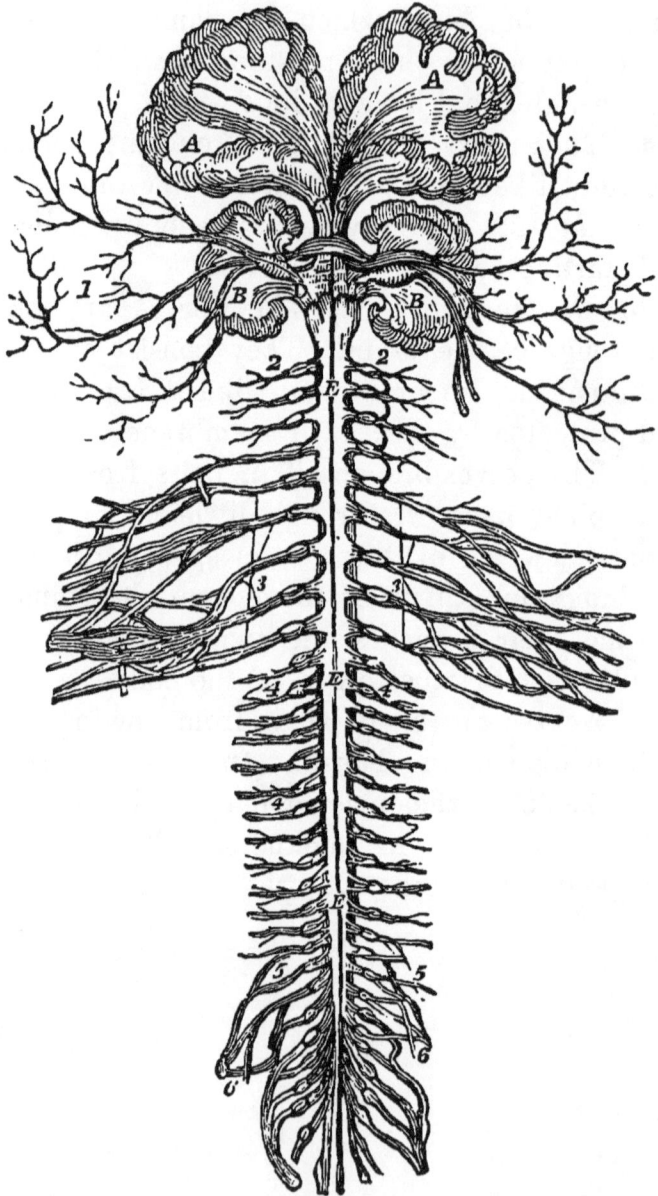

FIG. 32.

EXPLANATION OF FIG. 32.

A, A, the *cerebrum.*
B, B, the *cerebellum.*
C, C, the union of the fibres of the cerebrum.
D, D, the union of the fibres of the two sides of the cerebellum.
E, E, E, the spinal cord.
1, 1, the cranial nerves.
2, 2, the branches of the spinal nerves that extend to the neck and organs of the chest.

3, 3, the branches of the spinal nerves that extend to the arms and fingers.
4, 4, 4, 4, the dorsal nerves that extend to the walls of the chest, back, loins, and abdomen.
5, 5, the lumbar nerves that also extend to the chest and abdomen.
6, 6, the sacral nerves that unite, and form the great sciatic nerve of the legs.

Lesson III.

THE NERVOUS SYSTEM.

Work.—1. The nervous system has distinct offices to perform. While one portion (the brain) is engaged in thinking, and in receiving pleasant or painful sensations, or in sending out its commands to the body, another portion (the nerves) is engaged in conveying information and in carrying orders to the different organs.

2. The nerves are divided into two classes; viz., the *sensory* nerves, and the *motor* nerves. The sensory nerves are connected with the organs of taste, smell, hearing, sight, and touch. They carry impressions to the mind of the effects produced upon them in these organs. The *motor* nerves are connected with the muscles. When the brain wills that a

muscle shall move, a message with power is sent to
that muscle through its *motor* nerve, and it moves.

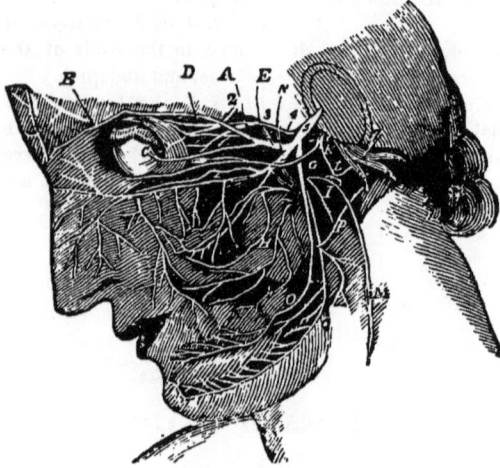

FIG. 33.

EXPLANATION OF FIG. 33.

2, the *optic nerve*, nerve of sight, connected with the eyeballs.

3, the *motor oculi*, used to move the eyes.

4, the *trochlearis*, which rolls the eye downward.

5, the *tri-gemini*, whose three branches extend to the upper part of the face,
to the upper jaw and teeth, to the lower jaw and teeth (this nerve is affected
in toothache), to the tear-gland of the eye, and to the nose.

O, the nerve of the tongue and of taste.

P, a branch of the nerve of taste, going to the ear.

Q, the nerve of the teeth of the under jaw, which finally comes out on the chin
to supply the muscles of expression.

7, the *auditory nerve*, being the nerve of hearing.

3. The nerves of sensation and of motion start
from different portions of the spine, but become
united in the same sheath soon after they leave it,
and till they enter the muscles. Thus every muscle
is moved by a *nerve of motion*, while beside it, in the

same sheath, is the *nerve of sensation.* If the mind wills that a finger be placed on any thing, the *motor* nerve moves the muscles of the finger, and the *sensory* nerve instantly reports to the brain whether that thing is cold or hot, rough or smooth. So when we smell, taste, or see any thing, or hear a sound, the nerves of sensation tell the brain whether it is sweet or sour, red or white, loud or low, etc.[1]

4. The nervous system is like a great telegraphic system. The brain is the great central office which receives and sends messages, and the nerves are the wires through which the messages are sent back and forth. If a nerve or a wire be severed, communication instantly ceases.

Fig. 33 gives a general idea of the nerves of the face.

Lesson IV.

EXERCISE OF THE BRAIN AND NERVES.

(a) *Exercise Beneficial.* — **1.** The brain and nerves suffer from lack of exercise, just as the muscles do.

2. Proper exercise of the nerves of motion relieves, or prevents, that distressing sensitiveness of the sensory nerves known as "nervousness."

-----

[1] All nerves of the muscles have a tendency toward the surface of the body.

3. The strength and activity of the muscles depend greatly upon the impulse given them by the brain and nerves. If the mind be pleasantly employed, the muscles will work long and actively without fatigue; but, if the mind be gloomy or inactive, the muscles soon grow tired.

(b) *Harmful Exercise.*—**1.** If the nerves of sensation be much exercised while the nerves of motion be but little, the former will be weakened by *too much* work, and the latter by *too little*.

2. Whenever the brain is over-exercised by hard study, or by excessive emotion or care, the blood rushes to it in increased quantities to replace the worn-out material : the veins and arteries become swollen, and a feeling of fulness or pain is caused. The over-exercised brain may in this way become diseased or paralyzed. In short, *well-regulated exercise* strengthens the faculties of the mind, while *inactivity*, or *injudicious exercise*, weakens them.

(c) *Equal Development.*—**1.** An *equal* development of all portions of the brain by a proper exercise of all the faculties of the mind and body is conducive to health and happiness.

QUESTIONS

EXAMINATION AND REVIEW.

QUESTIONS.

THE BRAIN AND NERVES.

Lesson I.

(a) — 1. Where is the brain situated? How surrounded and protected?

    2. Into how many and what parts is the brain divided?
How are the parts separated?

(b) — 1. What is first seen when the bones are removed? What
is the name of this membrane, and what is its office?
What lies next below the dura mater? Describe the
arachnoides. Just how is this coat placed? What is
the name of the inner coat? Describe it.

    2. Of what does the substance of the brain consist? Where
is the gray matter? The white? What is said of the
softness of the substance of the brain?

    3. What is the form of the outer surface of the brain?

(c) — 1. Of what is the brain the seat? How much is known of
its uses and work? *Note.* What is said of most of the
theories concerning its functions?

    2. What is believed to be the office of the cerebrum?

    3. What work has been attributed to the cerebellum? What
is generally admitted concerning its office?

(d) — 1. What is said of the insensibility of the brain in regard to pain ?

2. What effect has the removal of the upper part of the cerebrum of an animal?

3. What is said of the delicacy of the medulla oblongata?

Lesson II.

(a) — 1. From what do the nerves spring? To what do they extend ?

2. State what is said of the cranial nerves.

3. What is the spinal cord? Through what does it extend? What does it send out? What are these branches called ? To what do they extend?

4. In what manner do the nerves branch out from the spinal cord ?

(b) — 1. What are nerves ? Of what do they consist? How surrounded?

2. What of pairs of nerves? How many pairs spring from the brain, and how many from the spinal cord?

3. What is said of the size of nerves? What constitutes the *nervous system ?*

Lesson III.

1. State what offices the nervous system has to perform. What is the office of each part?

2. Into what classes are the nerves divided ? With what are the sensory nerves connected ? What is their work ? With what are the motor nerves connected, and how do they act?

3. From what portions of the spine do these classes of nerves start out? What is their position soon after leaving the spinal cord? With what is every muscle provided? Describe the action of the motor nerve. Of the sensory nerve. *Note.* What tendency have the nerves of the muscles ?

4. What is the nervous system like ? Explain why. What is the effect of severing a nerve or wire?

Lesson IV.

(a) — 1. What effect has lack of exercise on the brain and nerves?
    2. What does exercise of the motor nerves relieve?
    3. Upon what do the strength and activity of the muscles greatly depend? What is the effect of pleasant occupation of the mind? Of gloominess?

(b) — 1. What is the effect of great exercise of the sensory nerves while the motor nerves have but little?
    2. What takes place in the circulation when the brain is overworked, etc.? What may this result in?

(c) — 1. What is said of an equal development of all portions of the brain?

PART VII.

EYE, EAR, AND SKIN.

" The beams of light had been in vain display'd,
Had not the eye been fit for vision made.

    .   .   .   .

The watchful sentinels at ev'ry gate,
At every passage to the senses, wait;
Still travel to and fro the nervous way,
And their impressions to the brain convey."

EYE, EAR, AND SKIN.

THE EYE.

Lesson I.

THE EYE.

(a) *Position.* — **1.** The *eye*, the organ of sight, is situated in the upper part of the front of the skull, in hollows of the bones. It is surrounded, and protected from blows and accidents, by the bones of the socket in which it is placed.

(b) *Construction.* — **1.** The eyeball is surrounded by three coats; viz., the *sclerotic* or outer coat, the *choroid* or middle coat, and the *retina* or inmost coat. These coats lie one within another, like the layers of an onion, and hold the humors in globular shape.

2. *Nature of the coats.* The *sclerotic* (that is, *hard*) coat, like the dura mater of the brain, is thick, strong, and not sensitive. It has an opening in front, in which the cornea is placed. This coat gives great security to the delicate portions of the

161

eye, and affords attachment for the muscles. The *choroid* coat (*choroides*, fleecy) is very fleecy and

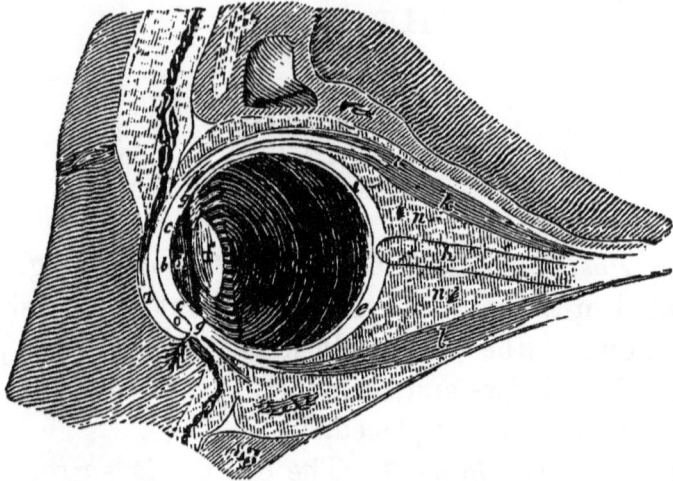

FIG. 34.

EXPLANATION OF FIG. 34.

This figure represents a section of the eye.

a, the upper eyelid, shut.

b, the *cornea*.

c, c, the cut edges of the *iris*.

d, the *pupil*, or round hole in the centre of the iris. In the real eye this looks like a bright black dot.

e, e, the cut edges of the *sclerotic* coat, the *choroid* coat, and the *retina*, or inmost coat.

f, the *crystalline lens*.

h, the *optic* nerve.

i, the *levator* muscle, that raises the eyelid.

k, the upper straight muscle of the eye.

n, n, a section of the blood-vessels and nerves, with a quantity of fat, surrounding the optic nerve.

The small dark space at *d* is occupied by the *aqueous* humor in the front of the eye: the large dark space back of *f* is occupied by the *vitreous* humor.

soft, and is composed of minute arteries and veins, which form a web about the eye. This accounts for the dark-red color of this coat. The *retina* (that is, *a net*) resembles ground glass in color, and is so very delicate that it cannot bear its own weight. It is really an extension and expansion of the optic nerve. It receives the rays of light, and is the immediate seat of sight.

3. *The cornea.* The *cornea* (*cornu*, a horn) covers the front of the eye and aqueous humor. In form and appearance it resembles a watch-crystal.. It is composed of thin, transparent plates, under the outermost of which are little sacks or glands, which give out an oily fluid that spreads over the surface, and gives this part of the eye great brilliancy.[1] When death approaches, this fluid collects in a dark cloud over the cornea.

4. *The iris.* The *iris* (that is, *the rainbow*) is that portion of the middle coat of the eye which lies back of the cornea. The coloring-matter of the eye is spread over its inner surface, black, blue, or brown, as the case may be. In the iris is a circular opening, called the *pupil* of the eye. The iris has the power of expanding and contracting, and thus enlarges or diminishes the size of the pupil (*pupilla*, a little puppet).

5. *The crystalline lens.* This "magnifying glass" of the eye is found between the two humors, just back of the pupil. It resembles a circular glass

---

[1] These little glands can only be seen by the aid of a powerful microscope.

button, convex on both sides. The crystalline lens
is held in place by a delicate, transparent envelope,
which connects it with the coats. It focuses the rays
of light.

6. *The humors.* The *aqueous humor* lies directly
back of the cornea, and fills the front chamber of
the eye. It is a perfectly clear, water-like fluid
(*aqueous*, like water). It sustains the cornea, and
keeps it always at the same distance from the pupil
of the eye. The *vitreous humor* (*vitreous*, glassy)
occupies the back chamber of the eye. It consists
of a substance like the uncooked "white" of an egg,
which is transparent, and allows light to pass through
it to the retina.

7. *The optic nerve.* This nerve springs from the
brain, passes through a bony canal, enters the back
of the eye, and branches off through the globe. The
small fibres of the nerve within the ball assume the
form of a web, and constitute the retina. The optic
nerve is about three-fourths of an inch long, and
somewhat larger than a straw.

8. *Glands.* The *lachrymal gland* (*lachryma*, a
tear) is a small sack in the upper and outer socket
of the eye, just above the ball. It prepares the tears,
and constantly pours out enough of its contents, by
pressure of the lids and rolling of the eye, to moisten
the surface of the eye, and prevent shrivelling. The
tears finally find their way to the inner corner of
the eye, and there enter little openings (*lachrymal
canals*), from which they flow into a bony tube

(*nasal canal*), and thence into the nose, whose inner surface they moisten.

EXPLANATION OF FIG. 35.

a, the *lachrymal gland.*
b, b, the eyelids, widely open.
c, c, the openings into the *lachrymal ducts.*
d, d, the *lachrymal ducts.*
g, g, h, the *nasal sack* and *duct.*

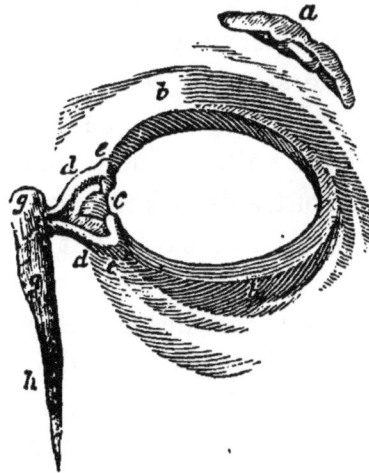

FIG. 35.

Lesson II.

WORK OF THE EYE, OR HOW WE SEE.

1. As yet, no one has been able to explain precisely *how* or *why* we see. We must await, from the progress of science, an explanation which physiology cannot now give.

2. We know that light is reflected from objects; that it enters the eye through the transparent cornea, passes through the aqueous humor, and enters the pupil; that it passes through the pupil, and reaches the crystalline lens, where its rays are bent from a

direct course. It is believed, that, after the rays reach the retina, a picture of the object is formed upon it, and that the impression is conveyed by the optic nerve to the brain, where the impression is understood or *seen*, but *how*, we do not know.

3. The iris expands and contracts independently of the will. When the quantity of light is too great, it contracts, diminishes the size of the hole in its centre, and shuts out some of the rays. When we leave a well-lighted room, and enter another where there is less light, the iris expands, and enlarges the pupil, in order to admit as many rays as possible. The pupil, therefore, is large or small according to the quantity of light necessary to make an impression on the retina.

Lesson III.

CARE OF THE EYE. — ABUSE. — DISEASE.

(a) *Care of the Eye.* — **1.** Care should be taken, in working or reading by lamp or gas light, *that the rays do not strike the eye directly.* The light should fall upon the work or the book, and not upon the eye. Allow the light to fall from above the level of the eye, or over the shoulder, but *do not face it.*

2. The nerves and muscles of the eye become fatigued by long-continued work, and may become *permanently weakened by lack of rest.* Care should be taken to give them a few minutes' rest occasionally, when they are employed in reading fine print, sewing, etc.

3. In work or study, the eye should not be brought *unnecessarily near the object that claims its attention.*

4. Care should be taken *not to employ the eye in deficient light* habitually or frequently. This practice weakens the nerves of the eye.

5. *The eye demands cleanliness,* and should be bathed to remove dust and impurities.

(b) ***Abuse and Disease*** — **1.** *Myopia* (near sightedness) is a very common disorder of the eye. Much of it is caused by bending the head over, and bringing the eye too near an object, as in reading, writing, sewing, etc. This habit causes the cornea and lens of the eye to adapt their form to suit the nearness of the object, and in time they become unable to adapt themselves to objects at a greater distance.

2. In myopia the convex form of the lens or cornea is greater than in a natural condition. Myopia may be relieved by exercising the eye in looking at distant objects, and by the use of double *concave* spectacles. As "prevention is better than cure," *do not induce the disorder by abusing the eye.*

3. *Presbyopia* (far-sightedness) is caused by the flattening of the cornea or the crystalline lens. In this condition the eye cannot see near objects distinctly. Presbyopia does not usually make itself felt till about the age of forty. *Convex* spectacles relieve far-sightedness.

4. Few persons can see equally well with either eye. This defect may be inherited, but it is more frequently caused by one-sided use of the eyes. In this way one eye is compelled to adapt its organs to focus the rays of light from an object at a shorter distance, and the other eye from the same object at a greater distance, from the retina. One eye thus becomes *myopic*, and the other *presbyopic.*

5. *Cataract* is a very frequent cause of blindness. In this disease of the eye, a thick, milky matter spreads itself over the cornea, and shuts out the rays of light. To remove a cataract, the surgeon uses a thin lancet, and cuts away a portion of the cornea. (*See Fig. 36.*)

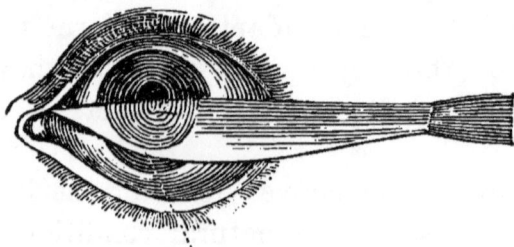

FIG. 36.

EXPLANATION OF FIG. 36.

This plan represents an eye surrounded by its natural appendages, with a knife passing through the anterior chamber. A dotted line indicates the lower edge of the flap, made by cutting off just one-half the cornea from its attachment with the sclerotica in order to allow the crystalline lens to escape whenever the knife is withdrawn.

6. The eye is *affected by the general health* of the body. Indigestion (*dyspepsia*) sometimes causes that troublesome affection of the eye commonly called "flying flies." Little motes, flies, or clouds appear to flit before the eyes. When the digestive organs return to a healthy condition, the motes disappear.

THE EAR.

Lesson IV.

THE EAR.

(a) *Location.*—1. The *ear*, the organ of hearing, consists of three parts; viz., the *external* ear, the *middle* ear, and the *internal* ear. The external ear is on the outside of the head, and the middle and internal portions are in the bones, at the base of the skull.

(b) *Construction.*—1. The *external* ear is a thin, elastic cartilage, concave on one surface, and convex on the other. Its concave surface consists of grooves which finally form one large basin at the entrance of the opening into the head. From the opening, a passage or tube, called the *auditory canal*, extends to the middle ear, or *drum*. This canal is about an inch in length, and its inner end is closed by a thin, tightly drawn membrane, called the *tympanic membrane*.

2. The *middle* ear (*tympanum*), or drum, is a small cavity which is separated from the auditory

canal by the tympanic membrane. The air within the drum communicates with the outside air by a passage called the *Eustachian tube*, which leads to the back of the mouth. Within the drum is a collection of four small bones, one joined to the extremity of another. From their shape, they have been named the *mallet*, the *anvil*, the *stirrup*, the *round bone*. *Fig. 37* represents these bones in their natural size, excepting the last one, which is magnified.

FIG. 37.

3. The *internal* ear, or *labyrinth*, consists of winding passages in the solid bone. The *auditory nerve* is spread over these passages like a lining, and they are filled with a watery liquid. One of these winding passages is named the *cochlea*, or snail-shell.

Lesson V.

WORK OF THE EAR. — CARE OF THE EAR.

(a) *How we hear.* — **1.** All things which produce sound vibrate in doing so, and communicate these quiverings to the air around them. The waves of air reach the external ear, which, like a funnel, re-

ceives as many of them as it can, and causes them
to flow along its channels into the auditory canal.
This canal conducts the air-waves inward, to the
membrane at its extremity.

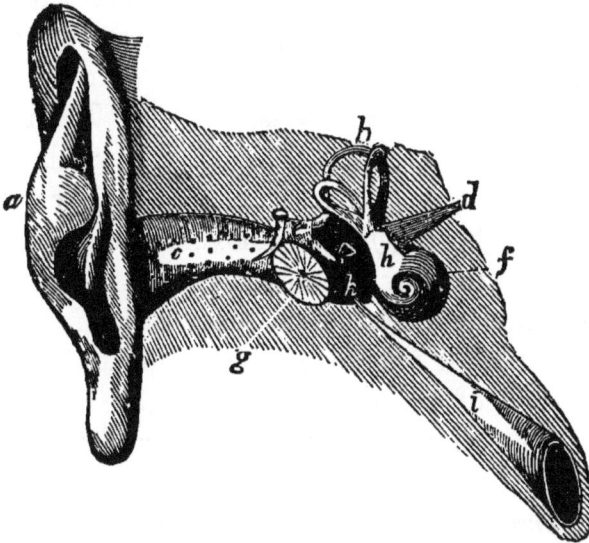

FIG. 38.

EXPLANATION OF FIG. 38.

a, the external ear.
b, the canals of the labyrinth.
c, the auditory canal.
e, the anvil-bone.
f, the cochlea.

g, the tympanic membrane.
k, the middle ear (*tympanum*), in
which the little bones are placed.
i, the Eustachian tube.

2. The air-waves beat upon the membrane of the
drum, and cause it to vibrate just as the head of an
ordinary drum does when it is struck.[1] The vibrations of the membrane cause the air within the drum
(*tympanum*) to vibrate, and to set the little bones to

---

[1] The vibrating plates of the telephone imitate this membrane.

vibrating and swinging, at the same rate. All these shakes and vibrations produce similar ones in the watery liquid in the labyrinth, and these produce some kind of an impression on the *auditory nerve*, which lines the inner ear. This nerve carries the sensation to the brain, which recognizes it, we know not how, as *a sound.*

(b) *Care of the Ear.—Disease.—*1. The auditory canal of the ear sometimes becomes partially closed, and the membrane of the drum covered, by ear-wax.[1] This should be *carefully* removed, occasionally. It hardens, and impairs the hearing.

2. *Habitual picking of the ears with pins* or other hard instruments should not be indulged in, as painful affections of the ear may be caused by this practice. These scraping instruments irritate the canal, and injure the head of the drum.

3. *Blows on the ears, boxing the ears,* may rupture the membrane of the drum, and injure the hearing. The firing of cannon, and other loud sounds made close to the ear, may produce the same effect.

4. Hearing is injured, or deafness caused, by disease in other organs of the body. Scarlet-fever, small-pox, measles, etc., sometimes produce partial or total deafness.

---

[1] Ear-wax keeps the lining of the ear moist and pliable. It also protects it from insects, as it is certain death to them.

THE SKIN.

Lesson VI.

THE SKIN.

(a) *Location.* — **1.** The *skin* is the outer covering, or envelope, of the body.

(b) *Structure.* — **1.** The skin consists of two layers; viz., the outer, or *scarf skin*, and the inner, or *true skin.*

2. The *scarf skin* consists of layers of flat, transparent scales, which are constantly being cast off and renewed. The dandruff of the head, and the white scurf that deposits itself on the clothing, are portions of the worn-out scarf skin. This part of the skin has neither nerves nor blood-vessels, and, when cut or punctured, suffers no pain. It is very thick over those parts of the body that are exposed to friction in working. This is especially true of the palm of the hand and sole of the foot.

3. The *true skin* is a dense, thick membrane, consisting of strong fibres that are arranged like those of felt cloth. This part of the skin is filled with small blood-vessels, which give it a bright-pink color. Besides the blood-vessels, the true skin contains *nerves, lymphatic-tubes, oil-tubes,* and *perspiration-tubes.*

4. The *arteries, veins,* and *capillaries* branch out all over the skin in a fine network. The *nerves* are so numerous that a needle cannot pierce the skin without touching one of them. The *lymphatics* are little tubes which open outwardly, on the under surface of the scarf skin, while inwardly they connect with the veins. The *oil-tubes* are very abundant. Their mouths open upon the outer surface of the skin, and may be plainly seen at the edges of the eyelids.

(c) *Work of the Skin.*—**1.** The skin, being tough and elastic, protects the tender flesh from injury. It also serves as an outlet for much of the worn-out or waste matter of the body, some of which is carbonic acid, some of an oily nature, and much of it perspiration.

2. The *perspiration-tubes* gather up, from the capillaries, waste matter in the form of water, salts, acids, etc., and carry it to the surface of the skin. The little mouths of these tubes are so numerous, that more than three thousand of them have been counted in one square inch of the skin. The work of these tubes goes on constantly. When their action is much hastened, they pour out the perspiration in so large a quantity that it may be seen on the skin, and this is called *sensible perspiration.* When the tubes do not discharge so rapidly as to cause the fluid to be seen on the skin, it is called *insensible perspiration.*

3. The *oil-tubes* carry a kind of oil from the blood, and pour it over the skin, to keep it moist and pliable.

4. The *lymphatics* absorb substances from the surface of the skin, and carry them into the veins.

Lesson VII.

CARE OF THE SKIN. — DISEASES.

1. The skin cannot be made whiter, permanently, by *the use of preparations and cosmetics.* These finally roughen and injure the skin. The lymphatics absorb portions of the substances spread upon the skin, and disease may be caused thereby. The bath, exercise, and pure air are the best beautifiers of the skin.

2. *Frequent bathing* is required to remove impurities from the surface of the skin. The perspiration-tubes are constantly depositing portions of worn-out matter upon the surface of the skin. If this waste matter be not removed *by washing the skin of the entire body,* the pores become clogged, and the work of these cannot be well performed. Soap should be used to dissolve the oily matter which accumulates. If these impurities be left upon the skin, *they may be absorbed by the lymphatics, and carried back to poison the blood, and cause fever.*

3. *Corns* are a thickened, hardened portion of the skin, caused by long-continued pinching of the joints of the toes by tight shoes.

4. *Skin-worms* are merely hardened oil (*sebaceous matter*) which forms in the outer openings of the oil-tubes when these do not perform their work perfectly.

5. *Ringworm* is an eruption of the skin.

6. *Freckles* are thick collections of the coloring-matter, frequently seen in the skin of persons of fair complexion.

NOTE. — The coloring-matter of the skin is spread over the true skin. This gives the varieties of color seen in the blonde, the brunette, and in the different races of 'men.

QUESTIONS

FOR

EXAMINATION AND REVIEW.

QUESTIONS.

PART VII.

EYE, EAR, AND SKIN.

THE EYE.

Lesson I.

(a) — 1. Where is the eye located? What protects it?

(b) — 1. By how many coats is the eye surrounded? Name them. How are these coats arranged? What office do they perform?

2. What is the nature of the sclerotic coat? What opening has it? What are the uses of the sclerotic coat? What is the nature of the choroid coat? Of what is it composed? What is the nature and appearance of the retina? What in reality is it? What work does it do?

3. Where is the cornea placed? What does it resemble? Of what is it composed? What fluid does it send out? What is said of this fluid?

4. What is the iris? What is said of the coloring-matter of the eye? What opening has the iris? What power of motion has the iris?

5. Where is the crystalline lens located? What does it resemble? How held in place? What work does it perform?

6. Locate the aqueous humor. Describe it. What work does it perform? Locate the vitreous humor. Of what does it consist, and what is its use?

179

7. From what does the optic nerve spring? Describe its
   progress. What form do its small fibres assume within
   the ball? What is the size of the optic nerve?
8. What are the lachrymal glands? What work do they
   perform? Where do the tears finally go?

Lesson II.

1. Can the act of seeing be precisely explained?
2. What do we know of light and objects? How does light
   enter the eye? Through what does it then pass? After
   passing ·through the pupil, where does it go? What
   occurs to the rays in the crystalline lens? What is
   believed to occur when the rays reach the retina?
   How does the picture reach the brain?
3. What movements of the iris are mentioned? What occurs when the light is too great? What does the iris
   do when we leave a light room, and enter a dark one?
   What, then, is the size of the pupil?

Lesson III.

(a) — 1. What care of the eye should be taken in working or reading by lamp or gas light? How should the light fall?
2. What effects have long-continued work upon the nerves
   and muscles of the eye? What care should be taken
   to prevent fatigue?
3. What is said of bringing the eye unnecessarily near objects?
4. What is said of deficient light?
5. What is said of the eye and cleanliness?
(b) — 1. What is myopia? By what is it often caused? How does
   this habit cause near-sightedness?
2. What is the form of the lens in myopia of the eye? How
   may myopia be relieved? What is better than cure?
3. What is presbyopia, and what causes it? How is the sight
   affected? When does presbyopia begin to be felt?
   What relieves it?

4. Can all people see equally well with either eye? From what may this difficulty arise? Explain.
5. Describe cataract of the eye. How is cataract removed?
6. How does the health affect the eyesight? Dyspepsia?

THE EAR.

Lesson IV.

(a) — 1. Of what parts does the ear consist? Name them. Where are they situated?

(b) — 1. What is the external ear? Describe its concave surface. What is the auditory canal? Describe it.

2. What is the tympanum or middle ear? With what does the air within it communicate? What is the Eustachian tube? What is found within the drum? How are these bones arranged, and what are they called?

3. Of what does the internal ear consist? How is the auditory nerve disposed in these passages? With what are they filled? What is one of these passages named?

Lesson V.

(a) — 1. How do things which produce sound move? What do they communicate to the air around them? What becomes of the air-waves? How does the external ear receive them? Into what do they then pass, and to what?

2. How do the air-waves affect the head of the drum? *Note.* What is said of the telephone? What does the vibration of this membrane cause? What is the final effect of all this?

(b) — 1. With what does the tube of the ear become clogged? What should be done?

2. What is said of picking the ear with pins, etc.?

3. What is said of blows on the ear? Firing of cannon?

4. What is said of diseases and hearing?

THE SKIN.

Lesson VI.

(a) — **1.** What is the skin?

(b) — **1.** Of what does the skin consist?

2. Of what is the scarf skin composed? What is dandruff, etc.? Has the scarf skin blood-vessels and nerves? What of its thickness?

3. Describe the *true skin.* What does the true skin contain besides blood-vessels and nerves?

4. How are the arteries, veins, etc., arranged in the skin? How numerous are the nerves? What are the lymphatics? What of the oil-tubes, or *sebaceous* glands?

(c) — **1.** What does the skin protect? For what does it serve as an outlet? Of what does the cast-out matter consist?

2. What is the work of the perspiration-tubes? How numerous are they? What is sensible perspiration? Insensible?

3. What is the work of the oil-tubes?

4. What is the work of the lymphatics?

Lesson VII.

1. What is said of the use of cosmetics, etc.? What is the danger in their use? What are the best beautifiers of the skin?

2. Why is frequent bathing necessary? What is deposited on the skin by the perspiration-tubes? What would be the effect of leaving these impurities on the skin?

3. What are corns? How caused?

4. What are skin-worms?

5. What is ring-worm?

6. What are freckles?

PART VIII.

EFFECTS OF ALCOHOL

ON THE

HUMAN SYSTEM.

―――――

"Oh that men should put an enemy in their mouths to steal away their brains! that we should, with joy, pleasance, revel, and applause, transform ourselves into beasts." — SHAKSPEARE.

EFFECTS OF ALCOHOL ON THE HUMAN SYSTEM.

Lesson I.

ALCOHOL. — WHAT IT IS.

(a) ***How Alcohol is said to have been Discovered.***
— **1.** Until men began to study alchemy, the liquid called *alcohol* was not known. The alchemists were men who directed their study and labor toward. discovering two objects, — one to make gold, or change common metals into it; and the other to discover a substance, called the "elixir of life," which was to give perpetual youth and vigor, and prevent death in those who partook of it.

2. The desire to discover these wonderful things led the alchemists to make very many experiments, and it is said that Paracelsus, a distinguished alchemist, during his experiments discovered alcohol, and became acquainted with its exciting properties. Believing that it would give permanent strength, he eagerly used it himself, and induced others to follow

185

his example. After boasting that the liquid gave him assurance of great length of life, his early death was caused by a course of violence and intoxication.

(b) *Derivation of the Word "Alcohol."* — **1.** The word "alcohol" is derived from the Arabic *al-khol*, meaning the powder of antimony, a substance with which some of the natives of Asia stain their eyelids, and thus, as they imagine, increase their personal beauty.

2. As this powder is very fine and pure, the name which originally belonged to it was given, in course of time, by Europeans to the liquid known by us as *alcohol*. The Arabs never called the liquid by that name.

(c) *What Alcohol is.* — **1.** Alcohol is a clear, water-like liquid, of a hot, biting taste, and it has a slight and not unpleasant odor.

2. Alcohol is not formed by distillation: it exists in simple fermented liquors, from which it is merely separated by the still. It gives to the liquors known as brandy, rum, whiskey, gin, etc., their intoxicating properties.

3. It is not certain that the ancients were acquainted with stronger liquor than wine, which, when perfectly made from the pure juice of the grape, is certainly of great intoxicating power.

(d) *How Alcohol is formed.* — **1.** There is only one source from which alcohol is obtained; namely, the fermentation (Lat., *fermentum*, to boil) of sugar, or of substances containing sugary matter.

2. When the juice of apples is first pressed out it is sweet, and has none of the sharp taste of cider. It does not become cider until it has fermented, or "worked," which action takes place after the juice has stood for a time.

3. The juice of the apple, the grape, grain, or other vegetable from which alcohol is obtained, is composed mainly of sugar and water, flavored with the particular taste of the fruit or vegetable; but after fermentation the juice loses its sweet flavor, and a portion of it has been changed into alcohol. Neither the water nor the flavoring-matter has been changed: the sugar, only, has become alcohol.

(e) *What Fermentation is.* — 1. When the juices of the vegetable have been allowed to stand for a time, decomposition begins. Now, sugar and alcohol are composed of the same elements, only not in the same proportions. Each consists of carbon, oxygen, and hydrogen.

2. When fermentation sets in, bubbles filled with carbonic acid gas arise to the surface, and the gas escapes. In this way a portion of the carbon and some of the oxygen of the juice are set free; but the hydrogen remains. The carbon, oxygen, and hydrogen which still remain, form the liquid known to us as alcohol.

3. Sugar, then, is separated into two parts, namely, *carbonic acid gas*, which is allowed to escape, and *alcohol*, which remains in the liquid. This process constitutes *fermentation*.

Lesson II.

ALCOHOLIC LIQUORS.—USES AND NATURE OF ALCOHOL.

(a) *Kinds and Quantity of Alcohol.*—1. All intoxicating liquors contain alcohol, and it is this that makes people drunk. Brandy, whiskey, rum, and gin, which are called distilled liquors (Lat., *distillare*, to drop), are about one-half alcohol; port wine and sherry wine are about one-fourth alcohol; claret and the white wines about one-tenth; and beer and cider have still less. Men usually cause those that contain the least alcohol to have the same effect as those that contain most by drinking larger quantities of them.

(b) *Some Uses of Alcohol.* — 1. Alcohol is much used in medicine and in the arts. Medicines are frequently prepared by mixing drugs with it. Cologne and other perfumes are made by flavoring it with the different oils and essences; and varnishes are made by mixing gums and resins with it. When mixed with turpentine, it forms camphene and other dangerous burning-fluids.

2. Alcohol will not freeze; and therefore it is colored red, and used in thermometers instead of mercury.

3. Alcohol has a great liking for water, and readily mixes with or absorbs it. Meat put into alcohol will remain good for a long time; for the alcohol absorbs the watery portions, and thus prevents decay. For

this reason it is much used by doctors and others in preserving. the flesh of specimens. But we cannot pause to mention all of its better uses, for, when rightly used, alcohol is a valuable servant.

4. Man does not always use alcohol rightly, however. Instead of keeping it as an obedient servant, he makes it a terrible, merciless master.

(c) *Stimulant and Narcotic.* — **1.** Alcohol is both a stimulant and a narcotic when taken into the body.

2. As a stimulant, it excites the brain and nerves, hastens the circulation of the blood, and produces intoxication.

3. As a narcotic (*narkē*, stupor), it blunts the sensibility of the brain and nerves, and produces sleep or stupor. All narcotics, when taken in sufficient quantity, are poisonous, and produce death.

Lesson III.

STIMULANTS, ANCIENT AND MODERN.

(a) *Leaves and Roots.* — **1.** All races have acquired the use of stimulants in some form.

2. The Australian and other of the lower races of mankind use merely certain roots and leaves, chewing them for their strengthening qualities. This kind of stimulation is only one step beyond that which causes the lower animals to seek certain plants for medicine, when they do not feel well.

(b) *Advancement in Manufacturing Stimulants.*

— **1.** The next step in advance of procuring stimulants by chewing is that made by the agricultural races, who use the chief grain grown by them, which, when fermented, yields a stimulant.

2. Arrack is obtained from fermented rice, and is an exceedingly strong liquor manufactured in the East. This liquor probably reached Western Europe from Egypt, where it was very early known. It still forms the principal drink of African races.

3. The wandering or pastoral tribes used, and still use, the milk of their flocks and herds, mixed with the honey of wild bees, in making their fermented drinks. The vessels used were made of the skins of animals, which were also used for storing away wines in the East.

4. Various plants have been used in both civilized and uncivilized countries, for thousands of years, in making wines and liquors. Grape-juices, however, were formerly confined to the countries in the western part of Asia, in Egypt, Greece, and Rome. In China the use of wine was forbidden, and the vines were not allowed to grow. Mead, a drink made of water and honey, was used by the Scandinavians and Anglo-Saxons.

5. Alcohol, the latest product of the art of manufacturing stimulants, was not included among the drinks of the ancients, in any of its present forms, and was not known to savages until introduced by Europeans.

Lesson IV.

ALCOHOL AND DIGESTION.

(a) *Alcohol, and Appetite for Food.* — 1. Alcohol excites the stomach to quickened action, but does not give any considerable nourishment.

2. While the liquor continues to excite the stomach, there is no great desire for food, because the nerves which produce the sensation of hunger are affected by the alcohol so that they do not perform their natural work.

3. After the liquor has spent its force, the stomach and its nerves are left in a weak, partially paralyzed condition, and they do not crave food.

4. If alcohol be taken regularly in small quantities, it causes the stomach to gradually lose its natural tone. It then becomes dependent on the artificial stimulus of the liquor, rather than on the natural vigor afforded by food.   .

(b) *Alcohol Delays Digestion.* — 1. One of the principal elements of the gastric juice is *pepsin*. It . has already been stated that alcohol has a great liking for water, and when it enters the stomach, it absorbs some of the watery portion of the gastric juice, and causes the pepsin to thicken, or coagulate. The pepsin then settles down in a fine white powder.[1] The

---

[1] Take the stomach of a calf or pig that has just been killed, rinse the gastric juice out of it with a very small quantity of water.

gastric juice thus becomes decomposed, and its nature is so changed as to make it unable to deal with the food. As a consequence, food taken into the stomach is not prepared for the body, and part of it passes out of the stomach unchanged, causing irritation and inflammation wherever it goes.

2. Continued indulgence in alcoholic drinks nearly always results in diseases of the digestive organs, — dyspepsia, inflammation of the stomach, inflammation of the bowels, diarrhœa, etc. Whenever these organs fail to do their work properly, all other parts of the system become deranged and the whole body suffers.

(c) *Alcohol Inflames the Stomach, etc.* — 1. Dr. Albert Day, an authority of world-wide reputation, says, "There is no appearance, after death, more common in the confirmed drunkard, who perishes after a long continuance of this habit, than a state of chronic inflammation of the lining membrane of the stomach. In this condition the walls of the organ are sometimes considerably thickened, are covered in their interior with a network of vessels closely injected with blood, and may present more or less extensive traces of ulceration. The thickening of the coats of the stomach may proceed to such an extent as to interrupt the passage of the food, through mechanical impediment."

---

and put the juice in a small bottle. The liquid will be milk-white, and, if a little alcohol be poured into it, the white portion will settle to the bottom. This white sediment is the *pepsin*, without which the other portion of the gastric juice cannot dissolve the food.

Lesson V.

ALCOHOL AND THE CIRCULATION. — THE HEART, ETC.

(a) *Alcohol Hastens the Circulation.* — 1. The cause of this is found partly in the action of alcohol on the blood-vessels, and partly in its effect on the nerves which govern the action of the heart. If the number of beats of the heart in twenty-four hours is about 100,000, the effect of an ounce of pure alcohol is to increase the number of beats to about 104,000 in the same length of time. The larger the quantity of alcohol taken, the greater is the number of beats in a given time. The hastening of the action of the heart has caused alcohol to be called a *stimulant* (*stimulus*, a spur).

(b) *How Alcohol Injures the Heart.* — 1. By stimulating the nerves of the heart, alcohol changes its natural action, and causes it to beat with undue rapidity, thus overworking and weakening its muscular power.

2. After the stimulus has spent its force, the heart is exhausted, and does not beat as quickly as before the alcohol was taken, and thus fails to propel the blood with natural speed. In this condition it is unable to perform its ordinary work, much less to perform any unusual task, if it should be called on to do so.

3. Continued use of alcohol may overtask the

heart so much as to relax its muscular fibres, and cause enlargement of its cavities.[1]   In this condition it may suddenly lose power of contraction (become paralyzed), and death would instantly result.

4. *Alcohol by constant use causes a softening of the muscular substance of the heart, and fattens it.*   This process is called *fatty degeneration.*   The more a muscle is thus degenerated, the weaker it becomes, because its muscular substance grows less, while the fat increases.   When fatty degeneration takes place in the heart, its walls become so soft that a finger could be easily pushed through them, and in this condition an unusual effort of the heart often causes its rupture from side to side, ending in sudden death.

(c) *Alcohol Relaxes the Small Arteries. — What Results. — 1.* Alcohol exerts a paralyzing influence on the nerves which govern the action of the muscular fibres of the arteries.   When these nerves are paralyzed, they permit the muscles of the small arteries to relax, and in this way enlarge the size of these little blood-vessels.   They then become swollen with blood in every part of the body.   "Carried to its full extent, this becomes congestion."   Thus it will be seen that alcohol deranges the entire circulation of the blood, and leads to disease of the heart and other organs.

---

[1] In heart-disease it is more especially hurtful by quickening the beat, causing congestion in the capillaries, and irregular circulation, and thus mechanically inducing enlargement of the cavities. — Dr. T. K. CHAMBERS.

Lesson VI.

ALCOHOL AND THE BLOOD.

(a) *How Alcohol Enters the Blood.* — **1.** When alcohol enters the stomach, some of it is instantly absorbed into the blood through the coats of the blood-vessels, without awaiting the slower process of absorption by the lacteals in the intestines.

2. Carried by the circulation to the heart, and then to the lungs, all that does not escape in vapor by the breath goes back again to the heart to be sent with the blood to all parts of the body.

(b) *How Alcohol Affects the Blood.* — **1.** The microscope has enabled us to discover how quickly the elements of food, drugs, and poisons make their appearance in the blood, and to learn how it is affected by them.

2. Alcohol mingles with the *plasma* of the blood, deprives it of its richness, and thus overcomes its power to nourish the system.

3. The *corpuscles* are caused to contract so much as *to be unable to absorb the usual amount of oxygen in the lungs, or carry out carbonic acid from the blood in the capillaries.* The coloring-matter of the corpuscles dissolves, they become pale, and their shape changes greatly. Some throw out matter, which floats about in the fluid portion of the blood.

4. The loss of strength in the corpuscles is indicated by black specks of fatty matter, which, in all cases of disorder in the blood, are found in great numbers.

(c) *Alcohol Interferes with the Burning of Waste-Matter.* — 1. By affecting the size of the corpuscles, alcohol diminishes the supply of oxygen in the blood, and thus prevents the development of heat, and checks the burning, or *oxidation*, of the waste-matter in the capillaries.

2. Portions of this dead matter are not changed by the burning process into carbonic acid and vapor, and therefore are not cast out in these forms by the lungs, but remain in the blood, and make it impure, *poison it.*

3. Impurity of the blood manifests itself in eruptions and pustules of the skin, scurvy, and boils. The system endeavors in these ways to cast out the impurities which the lungs, pores of the skin, and other organs, are unable to deal with in time to prevent disease.

(d) *Effects of a Weakened Condition of the Blood.* — 1. An influence which weakens the blood by depriving it of its nourishing properties must, of necessity, result in withholding from the body that which feeds its organs, in checking growth, and in injury to life. A weak and impure condition of the nutritive fluid is probably the first step to the starving and weakening of the body which it feeds.

2. The celebrated Dr. Virchow says that "alcohol

poisons the blood, arrests the development of the corpuscles, and hastens their decay."

3. The learned Dr. T. K. Chambers asserts that "alcohol impoverishes the blood, and there is no surer road to that degeneration of muscular fibre so much to be feared."

4. Dr. Benjamin W. Richardson, who has spent many years in investigating the effects of alcohol, and whose reputation is world-wide, says, "On the minute blood-vessels — those vessels which form the terminals of the arteries, and in which the vital acts of nutrition, and production of animal heat and force, are carried on — alcohol produces a paralyzing effect: hence the flush of the face and hands which we observe in those who have partaken freely of wine. This flush extends to all parts of the brain, to the lungs, to the digestive organs. Carried to its full extent, it becomes a congestion, and, in those who are long habituated to excess of alcohol, the permanency of the congestion is seen in the discolored skin, and too often in the disorganization which is planted in the vital organs, the lungs, the liver, the kidneys, the brain."

5. Mental disease of every grade, from the mildest depression of spirits to the most furious craziness, may be caused by the collection in the blood of the waste particles of the body.

Lesson VII.

ALCOHOL AND THE BRAIN.

(a) *Alcohol Affects the Substance of the Brain.* —
1. Alcohol hastens the circulation so much as to
overcrowd the minute capillaries of the brain, thus
causing congestion of these blood-vessels, and creating
a pressure that interferes with their healthy action.
Epilepsy and apoplexy frequently result from such
pressure upon the blood-vessels in the brain.

2. Congestion caused by stimulation is often indicated by a milky fluid found deposited beneath the
pia mater (inner coat) of the brain, when affected by
alcohol.

3. Alcohol hardens albuminous (like the white of
an egg) substances with which it is mixed. As the
brain is composed in great part of albumen, it becomes hardened by the alcohol carried into it by the
circulation; and doctors often find it unnecessary to
soak a brain in alcohol before dissecting it, because
it is already sufficiently hardened to suit their purpose. This hardening of the substance of the brain,
even when slight, interferes with its work, and dulls
the operations of the mind. When the hardening is
great, death results.

(b) *Alcohol Accumulates in the Brain.* — **1.** The
brain is more affected by alcohol than any other
organ. The tendency of alcohol to accumulate in

the brain is twice as great as in the liver, and three times as great as in other organs.

2. When a person has taken an excessive dose of alcohol, death immediately occurs from the sudden shock caused by the rapid massing of alcohol in the brain. The nerve-centres are paralyzed, all action of the organs ceases, and death results.

(c) *Derangement of the Brain.*—**1.** Unhealthful qualities of the blood caused by alcohol affect the size, shape, and color of the cells of the brain, and consequently their action, or thought-producing power. The brain then loses control of the mind, and insanity is the result.

2. The destructive effect of alcohol on the powers of the mind, through its action on the substance of the brain, presents a sad picture. Whenever the brain is too much excited by prolonged mental action or by alcoholic stimulants, disorder of the mind follows. It is not to be supposed that insanity always results from these causes. Usually, after a fit of intoxication, the brain returns to nearly its natural condition, and the mind recovers its power; but sometimes the brain and mind do not return to their natural condition, the mind continues weak and irregular in its action, and the person becomes a lunatic.

3. While insanity does not always result from continuous use of alcohol, some form of mental weakness is caused by the habit. The memory fails, the imagination becomes dull, the judgment becomes weakened, and the mind is kept fretful, irritable, and

dissatisfied. These conditions exhibit a wide difference between the bright and well-disciplined mind
that obeys the will, understands and reasons correctly,
and the mind that is partly deranged by an influence
which interferes with the structure and work of the
brain, the seat of the mind. Dr. Richardson says,
" I really doubt if a man who has been through the
dead-drunk stage of alcohol is ever quite the same
healthy man he was before." This remark applies
.to the health of the mind, as well as to that of the
body.

4. "If, by reducing the balancing power of the
vessels which regulate the supply of the blood to my
brain, I permit a more rapid current of blood to feed
my brain, I may for a time think more rapidly, and
express myself with more apparent energy. It is
clear, however, that under such circumstances I do
but exhaust more quickly, require to be wound up
more frequently, and wear out more speedily." —DR.
RICHARDSON.

5. Although the imagination may sometimes seem
to be stimulated to great power and activity under
the momentary excitement of alcohol, still the imagination, judgment, and every other faculty of the
mind, in time, become injured or destroyed by it.

6. Unless a healthy brain is present to guide the
judgment, we cannot expect a true and sound opin-
ion, nor a correct action as a result.

Lesson VIII.

ALCOHOL, THE NERVES AND MUSCLES.

(a) *How Alcohol Affects the Nerve-Pulp.* — 1. When alcohol reaches the nerves by means of the blood which circulates to them, it absorbs much of the water contained in the nerve-pulp, and leaves it so hard and dry as to be spoiled for its proper office. The delicate substance of the nerves is the one that, with the brain, soonest becomes affected by alcohol.

2. An authority says, " Alcohol instantly contracts the extremity of the nerves it touches, and deprives them of sense and motion, destroying their use." While under the influence of alcohol, a man may grasp a hot iron and be severely burned, or receive wounds or other injuries, without feeling much pain at the time, because the nerves are so much paralyzed as to be unable to feel, or to convey sensations to the brain. The brain itself is so affected as to be unable to receive the alarm, if it could be made by the nerves. Dr. Richardson says, "I learned through experiment, step by step, that the true action of alcohol in a physiological point of view is to create paralysis of nervous power."

(b) *Alcohol and Muscular Movement.* — 1. The nerves govern the muscles in all their movements. When, however, the sensibility of the nerves is blunted by alcohol, they fail to perform their work regularly and perfectly.

2. If the quantity of alcohol taken be sufficient, its influence extends to the spinal cord, and thence to the nerves that control and direct the movements of the muscles. Some of the nerves being quite paralyzed, they are unable to convey the commands of the brain, they lose all control of the muscles to which they belong, and motion cannot be produced. Others convey messages and power to the muscles so irregularly as to cause them to contract too much or too little.

3. Some of the muscles of the legs contract too much, and carry the feet too far: again they contract too little, and the feet are not carried far enough, in this way causing great uncertainty of movement. Control of the muscles of the hands is lost in a similar way. In course of time this deranged condition of the nerves and muscles becomes fixed, and the skilled workman is forced to seek rougher employment, in which delicacy of touch and exactness of muscular movement are not so much required. The final effect of alcohol is to permanently weaken both nerves and muscles.

Lesson IX.

HOW ALCOHOL AFFECTS THE TEMPERATURE OF THE BODY.

(a) *How Warmth of the Body is Kept up.* — 1. Animal heat is kept up by the burning of the worn-out particles in all parts of the body in which blood circulates.

2. The worn-out particles enter the capillaries, where the oxygen in the blood meets them and unites with them. The union of the oxygen with the waste-matter kindles a slow fire, which burns the carbon and hydrogen, and forms carbonic acid and watery vapor.

(b) *Alcohol Reduces the Heat of the Body.* — 1. Alcohol, from its effect on the corpuscles which convey oxygen into the blood (*See Lesson IV. (b), par. 3*), lessens the burning of waste-matter, and diminishes the warmth of the body.

2. To deprive the blood of its proper supply of oxygen has precisely the same effect upon its burning of waste-matter and production of heat, that withholding the proper supply of air has on the fire of the stove. The fire burns freely, and produces heat in exact proportion to the amount of air supplied. As alcohol tends to diminish the supply of oxygen, its effect is to arrest the development of heat, and finally reduce the temperature of the body.

3. In regard to the popular idea that alcohol supports the animal temperature, Dr. Richardson says, "It will be borne in mind that I have described a flush from alcohol as the first effect of it in its first stage, when into the paralyzed vessels the larger volume of blood is poured. In that stage, that is to say, in the earlier part of it, I found an increase of temperature. This increase, however, was soon discovered to be nothing more than radiation from an enlarged surface of blood; a process, in fact, of rapid cooling, followed quickly by direct evidence of cooling. After this I found that through every subsequent stage of the alcoholic process, — the stage of excitement, of temporary paralysis of muscle, of narcotism, and deep intoxication, — the temperature was reduced in the most marked degree. I placed alcohol and cold side by side in experiment, and found that they ran together equally in fatal effect, and I determined, that, in death from alcohol, the great reduction of animal temperature is one of the most pressing causes of death.

4. Varied and particular experiment has proved, beyond possibility of a doubt, that instead of being *a producer of heat* in those who use it, and for that reason a food in that sense, *alcohol is a reducer of heat*, and for that reason is not a food in that sense.

(c) *Alcohol, and Exposure to Extreme Cold.*—
1. The arctic explorers, Capts. Ross and Parry, Dr. Kane, and others, discovered that alcohol did not keep out the cold, and that men who did not use it

endured exposure to severe cold much better than those who did.

2. "In nearly all the cases of death caused by exposure to cold that I have known or heard of, it was found on inquiry, that the persons so dying had taken some alcoholic drinks, not necessarily in large quantity, before going out into a low temperature; the effects produced being languor, drowsiness, inability or disinclination to walk, stupor, and finally death. So well is this bad effect known by people in the north-west of America and in Canada, that they will seldom take even a single glass of spirits when about to be exposed to extreme cold." — DR. JOHN RAE, in *Medical Journal.*

3. Tests have been made with thermometers adapted to the purpose, and it has been found that the first flush caused by alcohol raises the temperature of the blood about half of a degree, but that the temperature soon sinks two or three degrees below 98, which is its natural warmth.

Lesson X.

INTOXICATION BY ALCOHOL.

(a) *The Stage of Excitement.* — **1.** In ordinary intoxication by alcohol, the first effect is a feeling of well-being and good nature.

2. Gradually, as the influence of the stimulant

increases, the excitement takes the form of extreme gayety, noisy mirth, or great talkativeness. The blood surges through the system; the brain takes part in the general whirl, and for a short time is spurred to great activity. The face is flushed, the blood-vessels become swollen, and the eyes flash.

(b) *The Stage of Mental Weakness.* — 1. In this stage, the stimulant has spent its strength, and reaction begins. The memory begins to fail; the thoughts become confused, and cannot be fixed longer than an instant on any thing; the temper is easily aroused; self-control is nearly or altogether lost, and offence is quickly taken at real or fancied affronts. In this condition the person may commit crimes, or be guilty of violent deeds, at which he would be horrified while in his natural state.

(c) *The Stage of Muscular Weakness.* — 1. In this stage, the nerves lose control of the muscles, and the man staggers, reels, and is unable to stand erect. (*See Lesson VIII.* (*b*), *par. 1, 2, and 3.*) Muscles of the lower lip, the eyelids, and the lower limbs, fail first. At length the man falls powerless.

(d) *The Stage of Stupor.* — 1. In this the last stage of intoxication, the narcotic properties of alcohol do their work. The temperature of the body falls with the dying power of the stimulant, and the man sinks into insensibility.

2. After an indefinite number of hours, the victim awakens from his stupor, and usually suffers from great thirst, terrible pains in the head caused by

congestion of the blood-vessels of the brain, sickness of the stomach, and distressing weariness of
the nerves and muscles. This condition of discomfort of mind and body may last a number of days
before the victim recovers from it. Nature permits
no violation of her laws without exacting a penalty.

Lesson XI.

DELIRIUM-TREMENS.

(a) *Character of the Disease.* — 1. *Delirium-
tremens* is a disease caused by excessive use of alcoholic liquors. One who has it is afflicted with tremor
of the entire body, sleeplessness, and delirium. The
disease is one of the forms of insanity.

(b) *Immediate Causes.* — 1. While this disease
is sometimes caused by a single fit of intoxication in
persons of very nervous temperament, it is usually
caused by excessive and long-continued use of alcohol. At times, delirium-tremens sets in while a man
is still continuing the free use of liquor; but in most
instances, the disease occurs when the hard drinker
suddenly quits the use of liquor temporarily. In
such instances, the weakened brain and nerves feel
the loss of the prop to which they have been accus-
tomed, and become entirely deranged.

(c) *Condition of the Victim of Delirium-tremens.*
—1. The first symptoms of the disease are great

nervousness and restlessness. A sudden noise, the opening of a door, or the entrance of a visitor, startles and excites the victim. His tongue and hands become tremulous, and he cannot sleep. If he chances to doze for a moment, he is aroused by horrible dreams.

2. Delirium soon begins, and the victim mutters to himself, or talks wildly to those about him. He imagines that he is surrounded by frightful monsters, snakes, and other loathsome reptiles. He makes desperate endeavors to escape from these, or from some one, who, as he imagines, wishes to do him harm, or to kill him.

3. The delirium is most frequently of this frightful nature, but not always so. Sometimes the insane fancy of the victim takes a droll or ludicrous form, and he appears to be highly amused by the comical pictures of his fancy.

4. While he is not often dangerous, still, in his frantic efforts to escape an imaginary enemy or danger, he may commit murder, or take his own life.

5. The delirium continues till the victim dies from exhaustion, or until he sinks into a stupor from which he may awaken comparatively sensible.

(d) *General Results of Delirium-Tremens.* — 1. If the strength of the victim has not been too much wasted by long use of alcohol, delirium-tremens is seldom fatal. Those whose strength and general health have been broken down by great use of alcohol frequently die from delirium-tremens. In such cases, death is often very sudden. The victim falls in a faint, from which he never recovers, or sinks in a stupor which ends in death.

Lesson XII.

EFFECTS OF ALCOHOL ON PEOPLE OF DIFFERENT TEMPERAMENTS.

(a) *The Nervous Temperament.* — 1. People of nervous temperament have very active brains and sensitive nerves, are easily excited, and readily become depressed, or "low-spirited." .

2. The stimulating property of alcohol increases the natural excitability of the brain in people of this temperament, and leads them into great excesses. The excitable brain and nerves, which at best are very difficult of control, become entirely unmanageable under the influence of so powerful a stimulant; and a person of extremely nervous nature may, while excited by alcohol, be as ungovernable as the most violently insane. Crimes of the most horrible nature are frequently committed by people while the brain is frenzied by alcohol.

3. The depression of mind, "lowness of spirits," to which nervous people are so liable after periods of great nervous excitement, is greatly increased by the narcotic elements of alcohol, which paralyze the brain and nervous centres. The action of the brain

is disturbed, the mind is clouded, and the nerves are unstrung. While in this depressed mental condition, people often commit suicide to escape the tortures of mind to which they are subject.

4. The nervous temperament should avoid every thing that tends to cause an increase of nervous excitability, all stimulants and narcotics.

(b) *The Sanguine Temperament.* — **1.** In people of this temperament the organs of the blood are very active, the blood circulates freely, and all the powers of the body are strong, and easily excited.

2. As the circulation of the blood in persons of sanguine temperament is naturally very active, alcoholic stimulants cannot be indulged without fear of disease. In such persons, stimulation creates so great an increase of the already active circulation, as to overwork the heart and blood-vessels, and produce paralysis, or other disease of these organs.

3. In the sanguine, as in the nervous temperament, stimulants are unnecessary and injurious.

(c) *The Lymphatic Temperament.* — **1.** People of this temperament are commonly stout, fat, or inclined to fatness; their skin is soft, and flesh somewhat flabby; the muscles are small and weak, and the whole body lacks vigor. The mind is less active than in any of the other temperaments. The temper is calm, and not easily aroused. The circulation of the blood is not as rapid as in the sanguine and the nervous temperaments.

2. It might be supposed that stimulation would

benefit this temperament; and so it would, if it could be made constant, and if there were no re-action or depression after stimulation. It must be borne in mind, that, while alcohol stimulates at first, it narcotizes or stupefies at last.

3. When the force of the stimulant has spent itself, the alcohol exerts its power as a narcotic, and the inactivity of mind and body which pertains to this temperament becomes greater. Thus it will be seen that alcohol increases, rather than counteracts, the weaknesses of the lymphatic temperament. Alcoholic stimulation should be avoided by persons who wish to permanently arouse activity of mind and body. Certain kinds of food, such as beef, mutton, coffee, etc., give nourishment and permanent stimulation, without depressing re-action.

Lesson XIII.

ALCOHOL AND MORAL CHARACTER.

(a) *The Moral Feelings Blunted.* — **1.** Alcohol not only weakens the powers of the mind, but also dulls the moral feelings. A carelessness about right and wrong is gradually induced by its free use, and a path is thus opened which leads, step by step, to dishonesty and other forms of crime.

(b) *Dishonesty of Speech.* — **1.** It leads to the

violation of truth, either to conceal the fault of in-
temperance, or the errors committed while under its
influence.

(c) *Dishonesty in Regard to Property of Others.*
— 1. Honesty in respect to the property of others
is violated, either to obtain the means to gratify the
appetite for alcoholic liquors, to pay the expenses of
extravagant habits which often accompany intoxication, or to provide for the pinching wants which such
habits occasion, and which cannot be provided for,
because of the loss of property and employment.

(d) *Crime in General.* — 1. Volumes might be
filled with the accounts of thefts, assaults, riots, fire-
setting, and murders, committed by those who have
given themselves up to the bad influences of alcoholic
liquors.

(e) *Appetite for Alcohol may be Inherited.* — 1.
Some of the best medical authorities say that an
appetite for alcoholic liquors may be inherited, just
as people inherit such diseases as scrofula, gout, or
consumption, and that it obeys all the laws that govern such diseases as are inherited from parents. It
is also declared by excellent authority, that this diseased appetite may skip a generation, and appear again
in a succeeding one with all its former strength.

Conclusion.

Your own observation and reflection will enable
you to compare the results which flow from the use

of alcohol, with the results that attend a steady course of industry, prudence, and wise care of the body and mind.

Lesson XIV.

THE STORY BRIEFLY TOLD.

1. Alcohol is a stimulant and a narcotic.

2. Alcohol interferes with appetite for food.

3. Digestion is delayed and made imperfect by alcohol.

4. Disease of the stomach and organs of digestion is caused by alcohol.

5. Alcohol unduly hastens the circulation of the blood, and causes congestion of the blood-vessels.

6. Alcohol increases the work of the heart, and thereby exhausts its power.

7. Alcohol softens the muscular fibres of the heart, and weakens it by changing the fibres into fat.

8. Alcohol relaxes the small arteries, and unfits them for their work.

9. Alcohol weakens the plasma of the blood, and overcomes its nourishing properties.

10. The corpuscles of the blood are contracted by alcohol, their size and form are changed, and their capacity to supply oxygen, and remove carbonic acid, is diminished.

11. Alcohol interferes with the burning of waste-matter in the capillaries, and thus poisons the blood, and prevents it from feeding the body.

12. Alcohol congests the blood-vessels of the brain, and causes apoplexy.

13. The substance of the brain is hardened by alcohol, and its thought-producing power injured.

14. Alcohol collects in the brain, and causes paralysis and death.

15. Alcohol affects the size, shape, and color of the cells of the brain, and produces insanity.

16. Alcohol absorbs water from the nerves, and paralyzes their action.

17. Alcohol, by its effects on the nerves, interferes with and weakens muscular movements.

18. Alcohol diminishes the heat of the body, and makes it sensitive to severe cold. It is not a protection against cold.

19. Alcohol affects injuriously men of all the different temperaments.

20. Alcohol intoxicates.

21. Alcohol causes delirium-tremens, and leads to other forms of insanity.

22. Alcohol tends to injure the moral sense, and leads to crime.

23. Appetite for alcoholic liquors. may be inherited.

QUESTIONS

FOR

EXAMINATION AND REVIEW.

QUESTIONS.

PART VIII.

ALCOHOL AND THE HUMAN SYSTEM.

___

Lesson I.

(a) — 1. About when did alcohol become known? Who were the alchemists?
   2. What is related of Paracelsus?
(b) — 1. From what is the word "alcohol" derived?
   2. How did alcohol obtain its name?
(c) — 1. What is alcohol?
   2. In what is alcohol first found? What property does it give to the various liquors?
   3. Were the ancients acquainted with pure alcohol?
(d) — 1. From what source is alcohol obtained?
   2. When does the juice of apples become cider?
   3. Of what is the juice of the fruits, etc., composed? What occurs after the juices have fermented? What has been changed into alcohol by fermentation?
(e) — 1. What happens when the juices stand for a time? Of what elements are sugar and alcohol composed?
   2. When fermentation takes place, what happens to the juices? What are set free by fermentation? What elements form alcohol?
   3. What, then, constitutes fermentation?

217

Lesson II.

(a) — 1. What do all intoxicating liquors contain? What liquors
are about one-half alcohol? What about one-fourth?
What still less?

(b) — 1. What of the use of alcohol in medicine and the arts?

2. Why is alcohol used in thermometers?

3. What is said of the liking of alcohol for water? How
does it preserve meat? Why do doctors and others use
it in preserving specimens, etc.? When is alcohol a
valuable servant?

(c) — 1. What is the nature of alcohol when taken into the body?

2. What are its effects as a stimulant?

3. What are its effects as a narcotic? What is said of all
narcotics?

Lesson III.

(a) — 1. What of the use of stimulants among all races of men?

2. What do the Australian and other lower races use as
stimulants?

(b) — 1. How do agricultural races obtain their stimulants?

2. From what is *arrack* obtained? How did this liquor find
its way into Europe?

3. How do the wandering tribes obtain their stimulants?
What vessels were used to hold the liquor?

4. What have been used, and for how long, in making wines
and liquors? To what countries was grape-juice confined? What of China? What is mead, and by whom
was it used?

5. What of alcohol and the ancients? Alcohol and the
savages?

Lesson IV.

(a) — 1. How does alcohol affect the stomach? Does it give
nourishment?

2. What is said of the appetite while alcohol excites the
stomach? Why is there no great desire for food?

3. What is said of the appetite for food after alcohol has spent its force? What is then the condition of the stomach and its nerves?

4. What is the effect of small doses of alcohol taken regularly? On what does the stomach then depend?

(b) — 1. What is pepsin? What does alcohol absorb in the stomach? What effect has it on the pepsin? What is the condition of the gastric juice then? How does this affect the digestion of the food? What becomes of the undigested food?

2. What are the results of the use of alcohol in respect to diseases of the digestive organs? What effect has disease of these organs on other parts of the system?

(c) — 1. What does Dr. Day state of the inflammation of the stomach by alcohol? Of ulceration? Of the thickening of the coats of the stomach?

Lesson V.

(a) — 1. How does alcohol affect the circulation of the blood? In what is the cause of this found? How does alcohol affect the beats of the heart? What is the effect of still greater quantities of alcohol on the action of the heart? Why is alcohol called a *stimulant?*

(b) — 1. What effect has the increased action of the heart on its power?

2. What is the condition of the heart after alcohol has spent its force? What is the result of this exhaustion of the heart?

3. What effect on the muscular fibres of the heart may continuous use of alcohol have? What may this weak condition result in?

4. What is said of softening and fattening the heart? How does softening and fattening a muscle affect its power? Why? When the heart suffers fatty degeneration, what may happen? Describe the condition of such a heart.

(c) — 1. How does alcohol affect the nerves of the arteries? What effect has this on the small arteries? What is then the condition of these arteries? What is this condition of

these arteries when carried to its full extent? What,
then, is the effect of alcohol on the entire circulation,
and what follows ?

Lesson VI.

(a) — **1.** What occurs when alcohol enters the stomach?
  2. Whither is it carried by the circulation? What becomes
of a certain portion of it?
(b) — **1.** What has the microscope enabled us to discover in respect
to food, drugs, etc., and the blood?
  2. How does alcohol affect the plasma of the blood?
  3. How does alcohol affect the corpuscles? What results
from this contraction of the corpuscles? What other
effects does alcohol have on the corpuscles?
  4. How is loss of strength in the corpuscles indicated?
(c) — **1.** How does alcohol affect the burning of waste-matter in
the blood? Tell what is said of the supply of oxygen.
  2. What is the condition, then, of portions of the worn-out
matter? What effect is produced by worn-out matter
remaining in the blood?
  3. How does impurity of the blood show itself? What does
the system endeavor to do with the impure matter?
(d) — **1.** What are the effects of a weak condition of the blood?
To what is weakness of the blood a first step?
  2. What does Dr. Virchow say of the effects of alcohol on
the blood?
  3. What does Dr. Chambers say of alcohol and the blood?
  4. What does Dr. Richardson say of its effects on the small
blood-vessels? What are the effects of this paralysis
of the blood-vessels? To what does the flush extend?
How is this congestion indicated?
  5. What is said of diseases of the mind, and collections of
waste-matter in the blood?

Lesson VII.

(a) — **1.** How does alcohol affect the capillaries of the brain?
What is the effect of this over-crowding of these
vessels? What diseases result?

2. How is congestion indicated when caused by stimulation?

3. What effect has alcohol on albuminous substances? Of what is the brain composed? How does alcohol affect the substance of the brain? In what condition do doctors sometimes find a brain? What are the bad effects of this hardening of the brain?

(b) — 1. How is the brain affected by alcohol when compared with other organs? What tendency has alcohol to collect in the brain?

2. What causes sudden death when a great quantity of alcohol has been swallowed at one time? What of the nerve-centres?

(c) — 1. How do unhealthful qualities of the blood affect the brain-cells? What is the effect on the mind?

2. What results from prolonged brain-work and from great stimulation? Does insanity always result? Does the brain always recover its health?

3. If insanity does not always result, what does? What difference do these conditions of mind present when compared with a healthy mind? What does Dr. Richardson say of a man who has once been "dead drunk"?

4. What, then, are the general effects of increasing the flow of blood unduly?

5. What is the temporary effect of alcohol on the imagination? What are its final effects on the faculties of the mind?

6. What cannot be expected if a healthy brain is not present?

Lesson VIII.

(a) — 1. How does alcohol affect the nerve-pulp? What portions of the body soonest feel the influences of alcohol?

2. What does an authority say of the effects of alcohol on the nerves? What is said of the grasping of a hot iron? Of wounds, etc., while under the influence of alcohol? What does Dr. Richardson say that he learned by experiment in regard to alcohol and the nerves?

(b) — 1. What govern muscular movement? What when the nerves are injured by alcohol?

2. How is the spinal cord, etc., affected by a sufficient dose of alcohol? How is the control of the muscles affected? How do the muscles then act?

3. How are the muscles of the legs affected? Of the hands? How is a skilled workman affected in course of time? What, then, is the final effect of alcohol on the nerves and muscles?

Lesson IX.

(a) — 1. How is the heat of the body kept up?

2. What takes place in the capillaries? What does the union of oxygen and waste-matter produce?

(b) — 1. What effect does alcohol have on the temperature of the body? By what means?

2. What are the effects of depriving the waste-matter of oxygen? What effect does the deprivation of oxygen have on the fire of a stove? In what proportion does a fire burn freely?

3. How does alcohol affect the temperature at first? What is this slight increase really? By what is it quickly followed? How is the temperature affected in the latter stages of intoxication? What is one of the most pressing causes of death from alcohol?

4. What has varied experiment proved in respect to the reduction of the temperature by alcohol?

(c) — 1. What is said of the testimony of arctic explorers in regard to alcohol and cold?

2. What is said by Dr. Rae in regard to deaths from exposure to cold? What is said of the use of alcohol by people in the north-western part of America and in Canada?

3. State the results of tests made with a thermometer in ascertaining the temperature of the body.

Lesson X.

(a) — 1. What is the first effect of alcohol in ordinary intoxication?

2. What gradually takes place? How is the brain affected? The face, blood-vessels, and eyes?

(b) — 1. When does re-action begin? How is the memory affected? The thoughts? The temper? What may a person in this condition do?

(c) — 1. What is the third stage of intoxication? What occurs in this stage? What muscles fail first? What happens finally in this stage?

(d) — 1. What is the fourth stage? What elements of alcohol now do their work? Into what condition does the person now sink?

  2. How long does insensibility last? What is the person's condition when he awakens from the stupor? How long may this disordered condition last? What is said of Nature and her laws?

Lesson XI.

(a) — 1. What is delirium-tremens? How is the victim of this disease affected?

(b) — 1. How is this disease caused? When, at times, does it set in? In most instances? What do the nerves and brain then miss?

(c) — 1. What are the first signs of the disease? By what ordinary things may he be startled? What of his tongue and hands? How is his sleep affected?

  2. Describe the victim's condition after delirium begins.

  3. What is most frequently the nature of the delirium? What is sometimes its nature?

  4. Is the victim dangerous? What may he do?

  5. How long does the delirium continue?

(d) — 1. Under what circumstances is delirium-tremens not fatal usually? Who frequently die from delirium?

Lesson XII.

(a) — 1. Describe the nervous temperament.

  2. How does alcohol affect the excitability? How ungovernable may a nervous person become under the influence of alcohol? What of the commission of crime?

3. To what are nervous people very liable after being strongly excited? How does alcohol affect this depression of mind? What do nervous people often do while in this depressed state?

(b) — 1. Describe the sanguine temperament.
2. Why cannot stimulants be taken without injury? Explain how the circulation is affected in this temperament by alcohol.
3. Is stimulation necessary in this temperament?

(c) — 1. Describe the lymphatic temperament. What of the circulation of the blood in people of this temperament?
2. What might be supposed to be beneficial? Under what circumstances would stimulation benefit this temperament? What must be borne in mind in regard to the nature of alcohol? How does alcohol affect the lymphatic temperament? By whom, then, should stimulants be avoided? What is said of the stimulating nature of certain kinds of food?

Lesson XIII.

(a) — 1. How does alcohol affect the moral feelings? What carelessness is induced by it? To what does this lead?

(b) — 1. How does alcohol affect the character for truth?

(c)˙ — 1. How does alcohol affect honesty in regard to the property of others? What temptations to dishonesty does it create?

(d) — 1. What is said of alcohol and crime?

(e) — 1. What do authorities say of an inherited appetite for alcohol? What is said of a skipping by the diseased appetite?

Lesson XIV.

1. What is alcohol?
2. What of alcohol, and appetite for food?
3. Of alcohol and digestion?
4. Of alcohol and the stomach?
5. Of alcohol and the circulation?
6. Of the work of the heart?

7. Of the muscular fibres of the heart?
8. Of the small arteries?
9. Of the plasma of the blood?
10. Of the corpuscles of the blood?
11. Of the burning of waste-matter?
12. Of the blood-vessels of the brain?
13. Of the substance of the brain?
14. Of alcohol collecting in the brain?
15. Of the cells of the brain?
16. Of the absorption of water from the nerves?
17. Of the muscular movement?
18. Of the heat of the body?
19. Of alcohol and the temperaments?
20. Of intoxication?
21. Of delirium-tremens?
22. Of the moral character?
23. Of inherited appetite?

PART IX.

TOBACCO AND ITS EFFECTS.

TOBACCO AND ITS EFFECTS.

Lesson I.

HISTORY OF TOBACCO.

(a) *When First Known by Europeans.* — **1.** Until the discovery of America, this plant was unknown to Europeans. The sailors who accompanied Columbus noticed the natives puffing smoke from their mouths and nostrils, and soon learned that this arose from the smoking of the dried leaves of a plant.

2. A friar, Roman Pane, who accompanied Columbus on his second voyage, noticed that the natives used the dried and pulverized leaves as a purgative medicine, by snuffing it through tubes of cane. The Aztecs of Mexico smoked it in highly ornamented pipes of silver; and other natives formed the leaves into rolls, and smoked them as the more modern cigar is now smoked. It appears that the ancient American races used tobacco in all the modes in which it is now used by man.

(b) *Tobacco Introduced into Europe.* — **1.** In course of time a quantity of tobacco was brought to Portugal, and in 1560 Jean Nicot, the French ambassador, brought some of it with him to France. In

1586 Sir Walter Raleigh, who had become acquainted with its use, introduced it into England.

2. In the short period of thirty years after its introduction into England, its use had become so common, and such enormous sums were expended in obtaining it, that his Majesty, King James, in the quaint style which was natural to him, said, " It is a custome loathsome to the eye, hateful to the nose, harmful to the braine, dangerous to the lungs, and in the black, stinking fume thereof neerest resembling the horrible Stigian smoake of the pit that is bottomlesse."

(c) *Origin of the Name " Tobacco."* — **1.** Various accounts are given of the origin of the name of tobacco ; but the one most probable is, that it was so called from the Indian *tabacos*, a name applied by the Caribs to the pipe in which they smoked the leaves of the plant, and thence came to be applied to the plant itself.

Lesson II.

NATURE AND EFFECTS OF TOBACCO.

(a) *Nature of Tobacco.* — **1.** Botanists describe forty different species of the tobacco-plant, all of which are more or less remarkable for their poisonous, narcotic properties.

2. The poisonous nature of tobacco is mainly due to one of its elements called *nicotine* (after Jean

Nicot), a substance similar to morphia. Nicotine is a liquid of a dark-brown color and of a biting taste. When vaporized by heat in a close room, it gives out an odor so oppressive, that breathing becomes difficult, even if but a drop of it has been spilled.

3. Nicotine is a deadly poison. Experiments show that five drops of it placed on the tongue of a dog have been sufficient to produce death, while twelve drops caused death in as many minutes. Two drops placed on the tongue of a fowl caused death almost instantly. Children who have inhaled the odor of nicotine from old tobacco-pipes, or who have swallowed minute particles of it, have been thrown into convulsions, and death has sometimes resulted.

(b) *Tobacco as a Medicine.* — **1.** Tobacco produces remarkable effects on the system, whether it be taken into the stomach, or applied to portions of the body from which the skin has been removed. In the latter instance it is absorbed into the blood, and its use is attended with great danger, sometimes with death..

2. When taken into the stomach, it produces great nausea, and this effect has suggested its use as an emetic; but, as the danger of such use is very great, it is seldom attempted. It also acts as a purgative.

(c) *General Effects of the Ordinary Use of Tobacco.* — **1.** When introduced into the system in small quantities, by smoking, chewing, or snuffing, it acts as a narcotic, and produces, for the time, a calm feeling of mind and body, a state of mild stupor and repose. This condition changes to one of nervous restlessness and a general feeling of muscular weakness when its habitual use is temporarily interrupted. In this condition, the body and mind feel in need of stimulation, and there is danger that a resort to alcohol may be had. The use of alcohol is frequently induced by that of tobacco.

2. When excessviely used, or used by one unaccustomed to it, it causes dizziness, nausea, faintness, vomiting, and extreme weakness; that is, it poisons; and convulsions and death may ensue.

3. Ordinarily the poison of nicotine is introduced into the system by swallowing small quantities of tobacco-juice, by its absorption through the lining of the mouth, or by inhaling the fumes of tobacco when it is smoked.

4. Tobacco, like alcohol, and for nearly the same reasons, injures the brain, deranges the entire nervous system, spoils the appetite for wholesome food, lowers the life-forces, injures the lungs and heart, and depresses the spirits. When indulged in by young persons, it saps the foundations of health, and dwarfs the body and mind.

Conclusion.

Every one is responsible for the care of his body and of his mind; and he who intelligently cares for "the house in which we live," will add to his powers and pleasures, and induce length of life.

QUESTIONS

FOR

EXAMINATION AND REVIEW.

QUESTIONS.

TOBACCO AND ITS EFFECTS

Lesson I.

(a) — 1. When was tobacco first known by Europeans? What did Columbus' sailors notice?

2. What use of tobacco was noticed by Roman Pane? How did the Aztecs use it? How was it used by some of the natives? In general, how was it used by the American races?

(b) — 1. When and by whom was tobacco introduced into France? Into England? How soon did its use become common?

(c) — 1. State what is said of the origin of the name " tobacco."

Lesson II.

(a) — 1. How many species of tobacco are described by botanists? For what are all the species remarkable?

2. To what is the poisonous nature of tobacco mainly due? Describe nicotine. What is said of it when vaporized?

3. What is the nature of nicotine? State how deadly it is when applied to the tongues of animals. How have children been affected by the odor of nicotine?

(b) — 1. What are the effects of tobacco when used as a medicine? When applied to the surface of the body?

2. When taken into the stomach? What use of it has been suggested by its effect on the stomach?

(c) — 1. What are its effects when introduced into the system in small quantities by smoking, etc.? How does this condition change? What danger is there?

2. What effect has the excessive use of tobacco?

3. How is the poison of nicotine ordinarily introduced into the system?

4. State the general bad effects of the use of tobacco. What are its bad effects upon young persons?

Conclusion.

What is said of our responsibility for the care of body and mind?

APPENDIX.

APPENDIX.

1. *Poisoning in General.* —In all cases of suspected poisoning, in which the kind of poison is not
known, the best thing to do is to cause vomiting.
Mix a dessert-spoonful of mustard, or a like quantity
of salt, with a tumblerful of tepid water, and give
it immediately. Cause the vomiting to continue till
the stomach has discharged all its contents, after
which, milk may be given freely. If the patient is
cold, warmth may be produced by hot tea or coffee.

2. *Poisoning by Carbonic Acid.* — A winter seldom passes without a number of deaths from the use
of coal or charcoal in close or unventilated apartments. The first thing to do is to remove the patient
from the poisonous atmosphere, and to open doors
and windows. Lay the person down, with his head
resting on his left arm. Open the mouth, draw the
tongue forward, and then roll the person gently
over toward the left, until the face is nearly downwards; then roll the body back again. The object

of this is to restore the breath by compressing the lungs, and then allowing them to expand again, and thus draw in the air. The body should be rolled as described about fifteen times a minute, and the process kept up for a long time; for persons have been restored even after an hour's effort. If the skin is warm, cold water may be poured on the head and spine; while, if the body be cold, a warm bath, or other means of warming, should be applied.

3. Restoration from Drowning.—Lay the person flat upon his back, and proceed precisely as directed in restoring the breath in cases of suffocation by carbonic acid. The operation of rolling the body should be kept up for a long time. Warmth should be applied to the body in any way that is most convenient.

4. Bleeding from an Artery.—Tie a handkerchief about the limb, *between the cut and the body.* Let the knot press upon the artery, and insert a stick in the folds of the bandage, and twist so tightly that the blood cannot flow from the compressed blood-vessels.

5. A knowledge of these few simple processes *may save a life* that would otherwise be lost ere the doctor arrives.

QUESTIONS.

APPENDIX.

1. What remedy should be given in all cases of suspected poisoning? How long should vomiting be kept up? What may then be given? What if the patient be cold?

2. What frequently results from the use of coal and charcoal? What is the first thing to be done? How should the suffocated person be placed? How should he be rolled? How often, and how long? What is the object of rolling the body? What may be done if the body is cold? If warm?

3. Describe the process of restoring animation in cases of apparent drowning.

4. How may bleeding from an artery be checked? Where should the bandage be placed? Why?

5. Of what use may a knowledge of these simple processes be?

Pronunciation and Derivation of Terms used.

KEY TO PRONUNCIATION.

ā, ē, ī, ō, ū, ȳ, long, as in āle, ēve, Īce, ōld, ūse, flȳ.
ă, ĕ, ĭ, ŏ, ŭ, ȳ, short, as in făt, mĕt, ĭt, ŏdd, ŭs, cȳst.
à, ä, ą, as in àsk, ärm, ąll.
e, ç, as in eat, çell.
ē, ę, as in ērmine, ęight.

ġ, g̃, as in ġem, g̃et.
ꞑ, as in liꞑk.
ȯ, ô, as in sȯn, ôrder.
s̩, as in has̩.
eh = k, as in ehorus.
ph = f, as in phantom.

Ab-dō'men (Lat.), from *abdere*, to hide; and *omentum*, entrails.

Al-bū'men (Lat.), from *albus*, white.

Al'ehe-mȳ (-kē-mē), from Arabic *al-kama*, the substance or composition of things.

Al'eo-hŏl (Arabic), from *al-kohl*, a powder of antimony.

Al'ī-ment'a-rȳ (Lat.), from *alere*, to feed.

Ā-năt'o-mȳ (Gr.), from *ana*, up; and *temo*, to cut.

Ā-ôr'tà (Lat.), *aorta*, to lift, heave; Gr. *aorte*, to keep in air.

Ap-pā-rā'tŭs (Lat.), from *apparare*, to prepare.

Ā'que-oŭs (ā-kwe-ŭs) (Lat.), *aqua*, water.

Ăr'tĕr-ȳ (Gr.), *arteria*, to keep in air. The ancients believed that the arteries contained air.

Ăr-tie'ū-late (Lat.), *artus*, a joint.

Ăs-phȳx'ī-a (-fĭx'ē-a) (Gr.), *asphuxis*, depriving of pulse.

Au'dĭ-tō-rȳ (Lat.), *audio*, to hear.

Au'rĭ-ele (Lat.), *auris*, an ear.

Bī-eŭs'pĭds (Lat.), *bis*, two; and *cuspids*, points.

Bīle (Lat.), *bilis*, anger.

Bī'çĕps (Lat.), *bis*, two: and (Gr.), *cephalus*, a head.

Brŏn'ehī (Gr.), *bronchos*, the windpipe.

Căp′ĭl-lā-ries (-rēs) (Lat.), *capillus*, a hair.

Căr-bŏn′ĭe (Lat.), *carbo*, a coal.

Căr′pŭs (Gr.), *karpos*, the wrist.

Căr′tĭ-lăġe (Lat.), *cartilago*, gristle.

Çĕr′e-bĕl′lum (Lat.) (diminutive of *cerebrum*, brain), the little brain.

Çĕr′e-brŭm (Lat.), the brain.

Chō′roĭd (Gr.), *chorion*, skin, leather; and *eidos*, resemblance.

Chȳle (kīl) (Gr.), *chulos*, nutritious juice.

Chȳme (kĭm) (Gr.), *chymos*, grayish juice.

Clăv′ĭ-ele (klăv′ĭ-kl) (Lat.), *clavicula*, a little key.

Cŏn-ġĕs′tĭon (Lat.), *congestio*, gathering into a mass.

Côr′ne-â (Lat.), *cornu*, a horn.

Côr′pŭs-çle (kôr′pŭs-l) (Lat.), *corpus*, a body; *corpusculum*, a little body.

Crȳs′tăl-līne (Lat.), *crystallinus*, congealed like ice.

Cū′ti-ele (kū′ti-kl) (Lat.), *cuticula*, from *cutis*, the skin.

Dĭ-ġĕs′tĭon (Lat.), *digestio*, separation, dissolving.

Dĭ′a-phrăgm (-frăm) (Lat.), *dia*, through; and *phragma*, partition.

Dĭs′lo-eāte (Lat.), *dis*, from; and *locus*, place.

Dĭs-sĕct′ (Lat.), *dis*, away; and *secare*, to cut.

Dis-tĭll′ (Lat.) *de*, from; and *stillare*, to drop.

Dȳs-pĕp′sĭ-â (Lat.), *dys*, bad; and *pepsis*, digestion.

Dūet (Lat.), from *ducere*, to lead.

Dū-ō-dĕ′num (Lat.), *duodeni*, twelve each.

Dū′râ māter (Lat.), *durus*, hard; and *mater*, mother.

Dĕ-lĭr′ĭ-um trē′meus (Lat.), from *delirare*, to go out of the furrow, to wander in mind; and *tremere*, to tremble.

Ĕp′ĭ-glŏt′tĭs (Gr.), *epi*, upon; and *glotta*, the tongue.

Ēū-stā′chĭ-an (yū-stā′kĭ-an), from *Eustachi*, the name of a learned Italian physician, who discovered the tube.

Fĭb′ū-lâ (Lat.), a clasp.

Fĭ′bre (-bŭr) (Lat.), *fibra*, a thread.

Fĕr-mĕnt-ā′tion (Lat.), *fermentum*, to boil.

Fē′mŭr (Lat.), *femoris*, the thigh.

Fūmes (Lat.), *fumus*, vapor, smoke.

Func′tion (Lat.), *fungor*, I act.

Găs′trĭe (Gr.), *gaster*, the stomach.

Glănds (-dz) (Lat.), *glandis*, an acorn, a nut.

Glŏt'tĭs (Gr.), *glotta*, the tongue.

Hū'mŏr (Lat.), from *humere*, to be moist.

Hū'me-rŭs (Lat.), the shoulder.

Hy'drō-ġĕn (Gr.), *hydro*, water: and *geinomai*, I produce.

Hy'ġĭ-ēne' (Gr.), from *Hygeia*, the goddess of health.

Im-pŏv'er-ĭsh (Old French), *povere*, poor.

In-săl'ĭ-vā'tion (Lat.), from *in*, with; and *saliva*, spittle.

In-tĕs'tĭne (-tĭn) (Lat.), from *intus*, on the inside.

In-tŏx'ĭ-eāte (Lat.), from *toxicum*, poison.

In-vŏl'un ta-rў (Lat.), from *in*, not; and *volitum*, will.

In-nŏm-i-nā'tă (Lat.), from *in*, not; and *nomen*, name.

I'rĭs (Lat.), the rainbow.

Lăb'ў-rĭnth (lăb'ȧ-rĭnth) (Lat.), *labyrinthus*, full of windings.

Lăe'tē-al (Lat.), from *lactis*, milk.

Lăeh'ry-măl (lăk'rĭ-mal) (Lat.), *lachryma*, a tear.

Lym-phăt'ĭe (lĭm-făt'ik) (Lat.), *lympha*, a colorless flúid.

Lăr'yn̄x (-ĭnx) (Gr.), from *larngx*, a whistle.

Lĭg'ȧ-ment (Lat.), from *ligare*, to bind.

Măs-tĭ-cā'tĭon (Lat.), from *mastico*, I mash, I chew.

Mĕ-dŭl'lă (Lat.), *medulla*, marrow.

Mĕm'brāne (Lat.), from *membrana*, a delicate skin.

Mō'tor (Lat.), from *motum*, to move.

Mū'eoŭs (-kŭs) (Lat.), *mucous*, slime.

Mĕt'a-tär'sŭs (Gr.), *meta*, beyond; and *tarsus*, ankle.

Mĕt'a eăr'pŭs (Gr.), *meta*, beyond; and *karpos*, wrist.

Mĭ'ero-seope (Gr.), from *mikros*, small; and *skopeo*, I see.

Mŭs'çle (mŭs'sl) (Lat.), *mus*, a mouse; *musculus*, a little mouse.

Năr-eot'ĭe (Gr.), from *narke*, numbness, stupor.

Nā'şal (Lat.), from *nasus*, the nose.

Noŭr'ĭsh-ĭng (Lat.), from *nutritum*, feeding, supporting.

Oĕ-sŏph'a-ġŭs (ē-sŏf'a-ġŭs) (Gr.), from *oio*, to carry; and *phagein*, to eat.

Ŏr'ġăn (Lat.), from *organum*, an instrument.

Ox'y-ġen (Gr.), from *oxys*, sharp, acid; and *geinomai*, I produce.

Păl'ate (Lat.), from *palatum*, the roof of the mouth.

Păn'ere-as (Gr.), from *pan*, all; and *kreas*, flesh.

Păr'a-lўze (Gr.), from *para*, beside; and *lysis*, to loosen.

Pa-tĕl'lă (Lat.), a little dish. (From *patina*, a dish.)

Pĕl'vĭs (Lat.), a basin.

Pĕp'sĭn (Gr.), from *pepsis*, a digesting.

Pĕr'ĭ-eär'dĭ-ŭm (Gr.), from *peri*, about; and *kardia*, the heart.

Pha-lăn'ġes (Gr.), from *phalagx*, a rank.

Phăr'y̆ux (-ĭnx) (Gr.), from *pharugx*, the gullet.

Phy̆ş'l-ŏl'o-ġy (Gr.), from *phusis*, nature; and *logos*, a description.

Pī'a mā'tĕr (Lat.), from *pia*, tender; and *mater*, mother.

Plăş'mȧ (Lat.), *plasma*, anything formed.

Pleū'rȧ (Gr.), *pleura*, a rib, the side.

Pūl'mō-nā-ry̆ (Lat.), from *pulmo*, a lung.

Pûr'ga-tĭve (Lat.), from *purgare*, to make clean.

Py̆-lō'rŭs (Gr.), a gate, a door.

Rā'dĭ-ŭs (Lat.), a staff, a ray.

Rĕt'ĭ-nȧ (Lat.), from *rete*, a net; or *retineo*, I hold.

Sā'erum (Lat.), from *sacer*, sacred.

Sa-lī'vȧ (Lat.), *saliva*, spittle.

Seăp'ŭ-lȧ (Lat.), the shoulder-blade.

Sele-rŏt'ĭe (skle-rŏt'ĭk) (Lat.), *scleroticus*, hard.

Se-bā'ceoŭs (-shŭs) (Lat.), from *sebum*, tallow.

Sĕn'sō-ry (Lat.), from *sensum*, to perceive by the senses.

Skĕl'e-ton (Gr.), from *skello*, I make dry.

Spīne (Lat.), *spina*, a thorn.

Spōre (Lat.), *sporos*, a seed.

Stĕr'nŭm (Lat.), from *sterno*, to spread out, to flatten.

Stĭm'ŭ-lănt (Lat.), from *stimulare*, to spur on.

Stŏm'aeh (stŭm'ak) (Lat. *stumachus*, an opening.

Stū'pĕ-fy̆ (Lat.), from *stupere*, to be struck senseless.

Sūt'ūre (-yŭr) (Lat.), *sutura*, from *suere*, to sew.

Sy̆n-ō'vĭ-al (sĭn-o'vĭ-al) (Gr.), *syn*, with; and (Lat.), *ovum*, an egg.

Tĕn'dŏn (Lat.), from *tendo*, I stretch out.

Thō-răç'ĭe (thō-rās'ĭk) (Lat.), from *thorax*, the chest.

Thō'răx (Lat.), the chest.

Tĭb'ĭ-ȧ (Lat.), the shin-bone.

Trū'ehe-ȧ (trā'kē-ȧ) (Lat.), from *trachia*, rough.

Trī-eŭs'pĭd (Lat.), from *tri*, three; and *cuspis*, a point.

Ty̆m'pa-nŭm (Lat.), a kettle drum.

Ŭl'nȧ (Lat.), the elbow.

Vălve (Lat.), from *valva,* a folding-door.

Vein (Lat.), from *vena,* a blood-vessel.

Věn'tri-cle (Lat.), *ventriculus,* from *venter,* the belly.

Věr'te-brá (Lat.), from *vertere,* to turn.

Vī'brāte (Lat.), from *vibratum,* to set in motion to and fro.

Vit'rē-oŭs (Lat.), *vitreus,* from *vitrum,* glass.